T0146118

Place Names

*How They Define the World
—and More*

Richard R. Randall

The Scarecrow Press, Inc.
Lanham, Maryland, and London
2001

SCARECROW PRESS, INC.

Published in the United States of America
by Scarecrow Press, Inc.
4720 Boston Way, Lanham, Maryland 20706
www.scarecrowpress.com

4 Pleydell Gardens, Folkestone
Kent CT20 2DN, England

Copyright © 2001 by Richard R. Randall

All rights reserved. No part of this publication may be reproduced, stored in a retrieval system, or transmitted in any form or by any means, electronic, mechanical, photocopying, recording, or otherwise, without the prior permission of the publisher.

American Names by Stephen Vincent Benét, from *The Devil and Daniel Webster and Other Writings*, Penguin Books
Copyright © 1927 by Stephen Vincent Benét
Copyright renewed © 1955 by Rosemary Carr Benét
Reprinted by permission of Brandt & Brandt Literary Agents, Inc.

Geographical Fugue, by Ernst Toch
Copyright © 1950 (Renewed) EMI Mills Music, Inc.
All Rights Reserved, Used by Permission
Warner Bros. Publications U.S. Inc., Miami, FL 33014

British Library Cataloguing in Publication Information Available

Library of Congress Cataloging-in-Publication Data

Randall, Richard R.
  Place names: how they define the world—and more / Richard R. Randall.
    p. cm.
    ISBN: 978-0-8108-3906-9
    1. Names, Geographical. I. Title.
  G105.R36   2001   00-059531
  910'.3—dc21

♾ ™The paper used in this publication meets the minimum requirements of American National Standard for Information Sciences—Permanence of Paper for Printed Library Materials, ANSI/NISO Z39.48-1992.
Manufactured in the United States of America.

Contents

Preface

My interest in places and their names stems from my earliest years. Born in Toledo, Ohio, I was fascinated with a nearby river called Maumee and a former canal called Erie. I often visited relatives elsewhere in Ohio and in California at places whose names impressed me. My later years took me to Canada, Europe, Latin America, and the southwestern Pacific. References to the names in my diaries quickly generate memories of many places. Another place name had an influence: Mount Rainier in the state of Washington. My middle name is Rainier, and my lineage includes the British admiral Peter Rainier, for whom the mountain was named.

A Fulbright scholarship to Austria permitted me to carry out research for a Ph.D. in geography. My topic of study was the status of the Slovene minority in the southern part of that country with regard to efforts over the years to merge their territory with adjacent Yugoslavia, where their ethnic counterparts lived. My travels in both countries provided a fascinating perspective of how place names related to the cultural backgrounds of the populations. I then became an analyst with the Central Intelligence Agency (CIA) to produce geopolitical studies of southeastern Europe. Later I was the Washington representative for Rand McNally and Company, responsible for obtaining cartographic and related data from federal and other agencies to support the company's role as the world's foremost private map and atlas maker. In 1973, I was appointed the geographer of the Defense Mapping Agency (DMA) and the executive secretary of the U.S. Board on Geographic Names (BGN), the interagency body responsible for capturing correct names in the United States and the rest of the world to meet official U.S. requirements. This assignment, which lasted more than twenty years, required continuous contact with the nine U.S. agencies having members on the board and gave me detailed knowledge about their functions and their specific needs for accurate place names.

I also worked with bodies of the United Nations (U.N.) as a U.S. representative to develop and direct various international names programs. This function required me to collaborate with other countries during the

Cold War, including the Soviet Union and its allied countries to the extent it was feasible. I presented many papers to professional and general audiences, provided articles and book reviews for professional journals, gave interviews to the press, and wrote numerous reports to meet U.S. government requirements. With support from the Pan American Institute of Geography and History (PAIGH) and the DMA (renamed the National Imagery and Mapping Agency [NIMA] in 1996), I designed a two-week course in 1987 to teach methods of names standardization to representatives of Latin American countries. I gave most of the class lectures each year (except 1988) through 1992. The course continues under the leadership of BGN officers and local authorities as the only one of its kind in the world.

Upon retirement, I knew that while many books on names had been published, none covered the picture I had come to know in great detail. At this point, I began to research official files, personal records, and other sources. My continuing contacts with people in BGN and in other U.S. and foreign organizations assured me access to a vast library of current materials on place names. I recognize and appreciate the scholarly works that many individuals have created in the field of place names, or toponymy, and I refer to such materials as appropriate. Although my writing in many ways reflects their impressive contributions, it is not my goal to produce a book of an exclusively academic nature.

I am grateful to numerous people in the United States and other countries. Special thanks are due to Dr. Meredith F. Burrill, the executive secretary of BGN from 1943 to 1973 and my mentor and friend until his death in 1997. He shaped many policies that still affect the nature and function of BGN and United Nations names programs. In addition, I thank many people on the BGN staffs for U.S. and foreign names. They include Donald Orth, formerly the BGN executive secretary for domestic names, Roger Payne, the current BGN executive secretary and executive secretary for domestic names, Randall Flynn, the BGN executive secretary for foreign names, and Gerd Quinting, a scientific linguist at NIMA. I also express thanks for the support of the NIMA and its erstwhile Latin American body, the Inter-American Geodetic Survey. In addition, I am grateful to the PAIGH for helping me initiate and teach courses on names standardization in Latin America.

I am indebted also to names experts in other countries for their collaborative work on numerous occasions. In the United Kingdom are H. A. G. Lewis, for many years the chairman of the Permanent Committee on Geographical Names for British Official Use (PCGN), Patrick Geelan, former secretary of the PCGN, and Paul Woodman, the current secretary. The collaboration of Henri Dorion and Michael Smart in Canada was also valuable. Also significant were the close working relationships I had with leading names authorities of other nations, and with people in the U.N. Economic and Social Council (ECOSOC) responsible for organizing conferences and sessions to promote names standardization.

Kelsie Harder of the State University of New York at Binghamton, long an active member of the American Name Society, was most helpful in reviewing the text and providing me with valuable ideas as to content. I also thank most sincerely my wife, Patricia Spencer Randall, formerly the director of the Office of Communications at the National Institute of Allergy and Infectious Diseases, U.S. National Institutes of Health. She made many suggestions to improve my writing style and worked long hours as my editorial assistant. She also put up with me on the numerous occasions when I disappeared into my office. For these and other reasons, I dedicate this book to her.

Richard R. Randall
October 2000, Washington, D.C.

Introduction

Many books and articles deal with place names. Reference works list country and state names along with their origins and how they may have changed over time. Travel publications identify names of places for planning vacations and trips. Scholarly articles discuss the linguistic and etymological nature of names. Trade publications and academic works indicate how names have become part of our vocabulary. Articles in the press feature humorous, bizarre, and otherwise unusual names. They also deal with place names in various ways, often highlighting how new names caused by political or other action generate much confusion. This book addresses these topics while also providing details about how the United States, other countries, and the United Nations deal with place names.

Events of the past decade dramatically demonstrate the dynamic nature of place names and illustrate the value of a publication covering such a broad topic. The end of the Cold War produced a number of new countries with changed names, and the countries also gave different names to many of their features. When the news media began to report on events in such areas, readers often were perplexed when seeing names not familiar to them. One case was Ukraine. When that country—formerly called "The Ukraine"—became independent from the Soviet Union in 1991, it began to change upward of 90 percent of its names. The impact on followers of local and international news about places in that nation was predictable. In general, they would be confused until the nomenclature became familiar. Other elements of the former Soviet Union also have programs to change place names so they will reflect their national languages and will mirror their historical origins.

Changes also have occurred in other parts of the world. In India, Bombay is now Mumbai, and Madras is Chennai. Burma is now Myanmar. Cambodia is Kampuchea. Zaire is now the Democratic Republic of the Congo. Furthermore, as virtually all nations increasingly recognize cultural groups with different languages, alternative names for features may result. South Africa now recognizes eleven official languages, and efforts are under way to develop new names reflecting these languages. India has as many as sixty-five languages that may lead to new names in various areas. Names long accepted can also acquire new spellings and, as populations increase and form new settlements or towns, different names arise. New and changing names are a part of history, current and past.

Other elements affect the need to eliminate confusion. One major problem is that many features can have two or even more local names. In addition, the forms of names in any country can vary if procedures for recording information about names are inadequate. Having a name positioned at an incorrect location—on a map or a street sign—also presents problems. Furthermore, the name of a feature in one country will probably require a different spelling on maps or reference works produced in countries with other languages. For example, English speakers with a map showing Munich as a place in Germany may be puzzled when traveling in that country where road signs give directions to a place called München.

Despite the fact that some names on maps or in other publications may be out of date or incorrectly located, they have played important roles over the years. People have associated themselves and their life activities with places whose names became elements identifying them. Many names become everyday words whose original association with places is in many cases all but forgotten. Artists, poets, authors, and musicians also use names to relate their works with certain places. Agricultural and commercial products have been given names for associational purposes. The conclusion is that place names do much more than define our world.

Behind these facts is a dramatic story. That story is how place names—also called geographic names—serve many human needs at the personal, national, and international levels. The story also details how national and international organizations have attempted to assure consistent spelling and placement of names for many purposes. The requirement for accurate names goes back to a need to correct the confusion caused by faulty maps produced by early traders and explorers. Those maps often showed different names for identical places, and military and scientific expeditions were frustrated by having to rely on such products. Postal services, land surveys, and a host of other administrative and personal aspects of life were complicated when names on maps or in records were incorrectly identified or misnamed. Inaccurate or misplaced names of ports and shoreline features on coastal charts presented hazards to marine navigation. While there may be stories of how these factors affected many nations over the centuries, this book focuses principally on experiences of the United States.

More than 150 years ago, the United States realized that inaccurate names of natural features and settlements registered in official files and shown on federal or local maps had become a major handicap to the nation's growth. In 1890, the United States created the world's first agency to deal with place names, the U.S. Board on Geographic Names (BGN). The board created procedures that help eliminate the many difficulties caused by inconsistent or haphazard naming and assure that official U.S. documents carry current names accurately located. Such procedures are commonly defined as "standardization." Other nations have formed similar organizations, and in 1955 the United Nations began efforts to develop programs that all countries could apply. The work of national bodies and the United Nations to standardize names has been impressive. These efforts compose a significant part of this book.

The six sections of the book cover the background and roles of names, the dynamics characterizing them, efforts to standardize them, problems related to the Cold War, how U.S. diplomatic, military, and intelligence agencies rely on the BGN for accurate names, and examples of names that are interesting or that for various reasons have become unacceptable. The book also discusses factors that have prevented a complete resolution of problems affecting place names. These include political, linguistic, and cultural conditions, and the inability of numerous governments to enact and enforce standardization. Although the United States leads the world in names programs, it nevertheless has not resolved all the issues.

In his research, the author has consulted numerous sources, including published books, articles in professional journals, and materials produced by the United States, the United Nations, and other official national or international organizations. A considerable amount of information also is based on personal records he compiled while serving as the executive secretary of the board and its Foreign Names Committee and as the principal U.S. delegate to meetings of the United Nations and its Group of Experts on Geographical Names for much of the time of his employment between 1973 and 1993, and on continuing contacts with U.S. and foreign experts.

This book supplies e-mail and standard addresses and phone numbers of organizations and individuals that can provide information about place names in the United States and elsewhere in the world. This makes it of special value to individuals seeking information about names. With its heretofore undocumented information about the broad nature and function of place names and procedures for dealing with them, especially as related to the United States, the book has reference value as well as appealing to a broad readership.

The index does not include all of the several hundred place names in this book which, because they were selected mainly to illustrate how they relate to various subjects, have little or no other reference value. Names and other items in the index were chosen as items of significance.

Shortened Versions of Organization Names

When the official name of an organization, a political entity, or other body is cited several times in the book, the shortened form of the name is normally placed in parentheses immediately after the first citation. Thereafter, only the shortened form may appear. If the full name is cited at widely separated points, however, the full name may be given. This list defines the shortened forms used in this book. Most of the terms refer to organizations associated with the United States

LIST OF SHORTENED FORMS

ACAN—Advisory Committee on Antarctic Names (of the USBGN)
ACEF—Advisory Committee on Extraterrestrial Features (of the USBGN)
ACUF—Advisory Committee on Undersea Features (of the USBGN)
APC—Antarctic Place-names Committee (official U.K. body for naming features in Antarctica)
ATO—Antarctic Treaty Organization (an international body for nations involved in research related to Antarctica)
BGN—Board on Geographic Names or the U.S. Board on Geographic Names
CIA—Central Intelligence Agency
CPCGN—Canadian Permanent Committee on Geographical Names (in 2000, CPCGN became GNBC: Geographical Names Board of Canada)

DMA—Defense Mapping Agency (U.S. agency to produce military maps and charts; became part of NIMA in 1996)

DNC—Domestic Names Committee (of the U.S. Board on Geographic Names)

DOS—Department of State

ECOSOC—Economic and Social Council (component of United Nations responsible for U.N. programs on geographic names)

FNC—Foreign Names Committee (of the U.S. Board on Geographic Names)

GNBC—Geographical Names Board of Canada

ICOS—International Committee of Onomastic Sciences (dedicated to the study of names of all kinds)

IAU—International Astronomical Union

IHB—International Hydrographic Bureau (responsible for developing marine safety standards)

NSA—National Security Agency (a U.S. agency involved with various communications)

NATO—North Atlantic Treaty Organization

NIMA—National Imagery and Mapping Agency (U.S. agency to produce maps and charts for military and intelligence purposes, created in October 1996; provides support for BGN Foreign Names Committee, Advisory Committee on Undersea Features, and Advisory Committee on Extraterrestrial Features)

PAIGH—Pan American Institute of Geography and History

PCGN—Permanent Committee on Geographical Names for Official British Use

SCAR—Scientific Committee on Antarctic Research (part of the Antarctic Treaty Organization)

STANAG—Standardization Agreement of NATO (concerning place names on NATO maps, charts, and gazetteers)

UN or U.N.—United Nations

UNESCO—U.N. Educational, Scientific, and Cultural Organization

UNGEGN—U.N. Group of Experts on Geographical Names; also Group of Experts

USBGN—U.S. Board on Geographic Names (also BGN)

USGS—U.S. Geological Survey (U.S. agency responsible for geological, hydrographic, and environmental studies; includes the National Mapping Division, which produces topographic maps of the United States and provides support for BGN and its Domestic Names Committee and its Advisory Committee on Antarctic Names)

I

The Nature of a Place Name

This segment sets the stage by describing what place names are, reviewing factors related to their origins and functions, and noting how they affect us. Since earliest times, place names have been needed to describe the nature of our physical and human world. In addition to specific places of perhaps limited extent, such as mountains or cities with recognizable physical or administrative characteristics, there are places called "regions" that occupy large areas and may have different kinds of features or associated elements. Also noted is the requirement that place names be put on maps so map users can comprehend the nature of the depicted areas and identify other relevant factors. The text on this topic also defines the nature and basic roles of maps and similar products.

1

What Is a Place Name?

DEFINITION AND FUNCTION

From our earliest moments, we become aware of place names. Our relatives and friends live in towns or perhaps countries that are mentioned as we grow up. When names of places appear in headlines as part of current events, they get our attention. Popular TV quiz shows often ask participants the names of national capitals. Books discuss curious and comical names of places. Tourists look for names of towns or scenic areas as they plan trips. Linguists and scholars investigate the nature and history of place names. A casual search of dictionaries reveals how place names have become part of our everyday vocabulary. Military commanders need references to names for operational planning purposes. Many people have deep feelings about the names of their places of birth or their ancestors' places of birth. And national and international organizations have programs to make sure place names are correctly spelled and accurately located.

Place names are indispensable to many levels of human communications. Even the way they are pronounced can be of primary importance. This book, however, deals mainly with written or graphical elements. It is useful to note that from a linguistic point of view, speech is "language" and is acquired naturally, while writing is a learned activity.[1]

What makes a place name important? This book answers that question but first defines the term and its functions. A place name[2] is a word or series of words that identify physical and administrative features on the earth and physical features on sea floors or planets. In this way, they are like such nouns as hand, leg, shoulder, or other words that identify parts of the body. A place name, however, is more than just a word or words that describe a type of feature. Naming practices in some countries may exclude terms that identify types of features, and instead give only the specific name.[3] In the United States and many other countries, most place names have two words that have a broader purpose. One word identifies the feature (such as river) and the second word serves as a specific name to distinguish it from other similar features. The name "Potomac River" adjacent to Washington, D.C., is an example. Place names are, in fact, also similar to personal names. The last name identifies one's family, and the first name identifies one of several members of a family.

A principal type of physical feature is one formed by geological or climatological actions. On the earth's surface, features include continents, mountains, oceans, lakes, rivers, and other elements whose size, shape, and location are discernable and are deemed important enough to identify by names. Similar features are on sea floors and planetary surfaces and also may have names.

A second type of feature found only on the earth's surface is that created by human activity. It includes political or administrative divisions such as countries, provinces, districts, and cities. They too have distinguishable characteristics.

A third category includes a variety of small natural or vegetative features, for example, a sand dune, some rocks, a section of a beach, or a tree. Such features may be found in areas presently or previously inhabited by people who do not have a writing system but give names to commemorate an event, to honor a religious occurrence, or to serve another cultural purpose.[4]

The United States and many other countries apply place names mainly to the first two major types of features described above. Some nations, however, also use place names to identify roads, bridges, buildings, farms, parks, and other humanly made entities having defined characteristics. Their maps and other reference materials reflect such naming policies. Although international recommendations to "standardize" national naming practices began more than forty years ago, many nations follow differing practices with varying results.[5]

Experts commonly agree that a named feature must have a location, size, and areal limits that distinguish it from other features. A physical feature also generally has to be large enough to appear clearly on a topographic map at a scale of 1:25,000 or smaller. For areal features (such as a lake) and most linear features (such as a mountain ridge), the coordinates of a center point must be calculated before a name can be applied.[6] For a river, the mouth (where it empties into another feature such as a lake or the point where it enters another country) generally is considered its point of location, since a center spot otherwise can be difficult to find. Although feature locations must be known and most physical features are generally permanent, some can change their shape and position. A glacier could be named, yet its extent could diminish radically with time. Similarly, a river can change its course and the location of its mouth.

Maps are perhaps the most common vehicle for identifying place names.[7] A major requirement is to place the name so a user can see it easily and recognize the named feature and its limits. One issue is to decide how many names can appear on a map. As scales diminish and areas shown on maps decrease in size, smaller features and less important administrative units cannot be depicted or otherwise identified with names. Another concern is to print names on maps so they do not overlap or visually interfere with other features. Maps portraying too much information or "clutter" can frustrate users seeking locations and names of places. It is the function of the map designer or cartographer to decide the placement of names on maps.

The use of place names varies in different parts of the world. Where the population is rural and nonliterate or lives in a restricted area, residents may identify features with only a single word. When asked for the name of a local river, they may say only "river" because they know of no other bodies of water with river characteristics in the area. Elsewhere, local societies may give a large feature more than one name. For example, people living in rural areas at widely dispersed locations along a river may have different names for the same feature. Such variations of naming habits, while locally justifiable, pose difficulties for authorities attempting to conform to a generally accepted international policy that any feature should have only a single name.[8]

ORIGINS OF PLACE NAMES

Without question, place names were part of even the most primitive languages. Then, as now, humans needed to know the location and the nature of certain features. The earliest humans certainly used sounds, as well as accompanying gestures, to indicate the location of important places such as sources of water, hostile people, and edible plants. As written forms of speech developed and people expanded their travel and communications, the sounds identifying places evolved into words now known as place names. With the increasing development and spread of speech and writing, place names became integral elements of oral and written communications. Yet even today people without written languages refer to places by names. An interesting case is that the Temiar people of Malaysia use songs with names to transmit information recalling memories and events of places.[9]

Numerous books and articles describe the origins of place names and indicate that names given to places reflect a number of cultural factors. Other publications describe how political, cultural, and linguistic changes have led to different spellings of the names of many places. A major, if not the oldest, written source of names may be the Bible, which refers to some hundreds of places as sites of events dating upward of 3,000 or more years ago. As is the case today, knowing the locations of named places early in history enabled readers to understand the significance of events within their geographic, cultural, and historical settings. Without reference to place names, the history and the messages of the Bible would be difficult to comprehend. While names are found in many parts of the Bible, chapters 12–16 of Joshua may account for the greatest number of references.

The first biblical citation of names is in chapter 2 of Genesis. The King James Version provides these accounts: "And the Lord God planted a garden eastward in Eden" (Genesis 2:8). "A river went out of Eden . . . and from thence it was parted, and became into four heads" (Genesis 2:10). "The name of the first is Pison; that is it which compasseth the whole land of Havilah" (Genesis 2:11). "And the name of the second river is Gihon. . . . it . . . compasseth the whole land of Ethiopia" (Genesis 2:13). "And the name of the third river is Hiddekel . . . which goeth toward the east of Assyria. And the fourth river is Euphrates" (Genesis 2:14).

The leading names expert in Israel, Prof. Naftali Kadmon, notes that only two of these named features can be definitively traced to places known today.[10] The location of Ethiopia corresponds closely with that of the present country. The Euphrates River is now the river with the same name. Many other places cited elsewhere in the Bible can be located as either historical or current features. Current news refers to a number of such places in Israel or nearby countries.

In the region of the world generally referred to as the Middle East[11] efforts continue to preserve names to satisfy the linguistic and cultural preferences of countries that claim historical rights to the names, if not to the territories in which they are located. These efforts have generated degrees of dissension between Israel and its neighbors.[12]

CATEGORIES OF PLACE NAMES

Given the fact that names are the results of human practices, researchers have long investigated the circumstances behind name origins. Several writers have developed categories of names based on their origins or other factors that, as might be expected, have a degree of similarity. George R. Stewart, the author of the classic book, *Names on the Globe*, describes two major classes of names.[13] "Evolved" names originated at a primitive level to identify features probably of significance to local inhabitants. "Bestowed" names were given by someone as a conscious act of naming. Within these categories, Stewart also recognizes ten types of names that have more specific kinds of origins. (In the following list, the descriptions are applied by the author and relate mainly to places in the United States.

There are, of course, many additional examples of each type, as well as other systems that describe name categories.)

Associative. The feature has a geographic or natural association with another feature. For example, the state of Mississippi got its name from the Mississippi River. Another such case could result when a feature got its name from a nearby creature. A historian at a class on geographic names taught by the author in Brazil in 1992 referred to literature indicating one particular mountain had a name derived from a name indigenous people gave a butterfly common to the feature.

Commemorative. The name honors someone. For example, George Washington is commemorated by many places having his last name.

Commendatory. A name praises someone or something. For example, the name of the state of Pennsylvania reflects the work of William Penn, who encouraged settlement there.

Descriptive. The feature looks like something. For example, in Mexico, one side of a mountain resembles a dead person lying down and is called El Muerto, the dead one. Curiously, the opposite side has a different appearance and a different name, El Picacho, the peak.

Folk-etymological. Over time, a name had its spelling and meaning changed because ultimate namers did not comprehend such attributes of the original name. For example, Picketwire, Arizona, was earlier named by the French as *purgatoire*. Such a process accounts for many American names that were formed (and spelled) by English-speaking explorers and settlers according to what they heard.

Incident. A noteworthy event took place at or near the feature. For example, Reform, Alabama, was named after one or more religious services caused people to "reform."

Manufactured. Two or more names or words are put together as a new name. For example, Texarkana is a combination of letters taken from Texas, Arkansas, and Louisiana.

Mistake. Names that have been repeatedly misspelled or respelled with time. For example, in England, the Cambridge River is the current spelling of a name that had three different earlier versions.

Possessive. The feature is part of territory owned or controlled by someone. For example, Reston, a community in Virginia near Washington, D.C., was developed by a person with the initials R. E. S.

Shift names. The name of one feature is applied to another. For example, in Michigan, a water feature called Grand Rapids gave its name to a city that developed nearby.

Names in the "mistake" category can have unusual origins. Guides conducting tours in Mexico sometimes tell a story about the origin of the name Yucatan. They say that an early Spanish explorer in the area, now called the Yucatan Peninsula, asked a native what the area was called. The answer in the local language meant "I don't understand you," which the explorer heard as "Yucatan." He was said to have recorded that statement as the name. Another story concerns the name Canada. A Portuguese sailing captain, when seeing the barren northeast coast of what is now Canada, was said to have written in his log "Ça nada," which means "It is nothing." Somehow that expression was the same as the ultimate name.

Another category is "migrational." This includes names given to previously named features by people moving from their homelands to other areas where they partially or completely displaced native populations. Such new names may have been those of places in the former homelands or totally new ones that reflect the language or customs of an incoming culture. Vikings invading England brought numerous Viking names. In addition, the Viking presence influenced the spelling of existing names. Such names could fall under the category "bestowed," but the process was probably completely devoid of any official nature.

Still another type has had considerable impact in the past decade: "political." This category resembles the commendatory class, but there are two kinds. First, a new town may receive the name of a person or event deemed important by authorities. Second, and on a much broader scale, authorities of new regimes may rename existing towns, administrative divisions, and other features. The goals are either to eliminate references to older or no longer acceptable governments or to commemorate new concepts or political leaders, thereby promoting the validity of the current authority. In the late 1980s, names in Eastern Europe began to change in anticipation of the fall of the Soviet Union and its communist form of government. Names of numerous towns, provinces, and natural features were changed to precommunist names. Many of the reestablished names were those in use prior to the date such governments gained power. In some cases, the new names also reflect the local languages that were earlier displaced by Russian as the then-official language. This process could also be part of the "bestowed" classification.

The political influence on names is also found elsewhere. One example is Nicaragua. When the Sandanistas took power there, they replaced names honoring the former Somoza regime with names favored by the new procommunist authorities. With the fall of the Sandanista government in 1989, the new government began a renaming process that honored neither the Somoza nor the communist era. Since then, and now as well, names elsewhere in the world change as new nations are born and as new or different political and cultural views are registered. Such circumstances may call for the restoration and official approval of what some say were the earlier and acceptable place names.

Permanent or totally consistent place names are important for many aspects of local, regional, national, and global communications, but the human element will continue to generate change and accompanying confusion.

NOTES

1. Personal communication, June 29, 1999, from Kelsie Harder, longtime member of the American Name Society. It is true that linguists may refer to "language" as only a spoken phenomenon. This interpretation excludes any written form of a language, which may be described as a "writing system." This book generally follows the common practice of applying the word "language" to both the spoken and the written systems. One term or the other, however, may be used where such references are useful in order to understand a particular discussion.

2. Place names are also called geographic or geographical names, and toponyms. This book prefers the term "place name" but may also refer to other forms. Chapter 11, "The Terminology of Names," defines these and other terms in more detail.

3. Chapter 11 defines the two elements common to most U.S. place names: "specific" (the name used to identify the feature) and "generic" (the type of feature).

4. R. D. K. Herman, "The Aloha State: Place Names and the Anti-Conquest of Hawaii," *Annals of the Association of American Geographers*, March 1999, 76–102. Reference to M. K. Pukui, S. H. Elbert, and E. Mo'okini, *Place Names of Hawaii*, 2d ed. (Honolulu: University of Hawaii Press, 1974). Author cites such customs. Such naming practices are generally characteristic of nonliterate people. The article also describes how Western settlers applied their own names to many features, while ignoring existing names of small places.

5. In 1955, the United Nations began an effort that led to a program for nations to develop common practices for applying place names. This program is discussed in chapter 9, "The United Nations Joins the Effort."

6. Most atlases have indexes of names that include the coordinates of center points of features. A valuable source of names information is the gazetteer, a publication with alphabetic lists of names with coordinates and other locational information. These items are discussed in chapter 13, "Gazetteers."

7. Chapter 3, "Maps Say Little without Place Names," presents a more detailed discussion of maps and names.

8. The United Nations issued this policy as a desirable practice. Its advantage is recognizable, but there is growing evidence that some communities—especially indigenous ones—want their local names and so-called official nomenclature to be equally valid.

9. Marina Roseman, "Singers of the Landscape," *American Anthropologist* 100, no. 1 (1998): 106–121.

10. Correspondence from Professor Naftali Kadmon, Hebrew University of Jerusalem, February 3, 1995.

11. Chapter 2, "Regional Names," discusses this and other regional terms.

12. Chapter 6, "Where in the World Is That Place?" discusses cases of controversy about such names.

13. George R. Stewart, *Names on the Globe* (New York: Oxford University Press, 1975), 74–84.

2

Regional Names

INTERNATIONAL REGIONAL NAMES

Regional names are another type of place name, although they may not be definable in the same way. Their locations and characteristics may be inexact, their definitions varying with time and according to those who define them. Nevertheless, they are important to many aspects of human communications. There are several categories of regional names. International regional names include the Alps, the Caribbean, Eurasia, Melanesia, Micronesia, Polynesia, Oceania, and the Pacific Rim. Because of geographic proximity or historical associations, their collective names are important identifiers. Regional names also may define groupings of several countries that have a geographic relationship. The names originated because of a need to refer to two or more countries associated with given areas. Typical names are North America, Central America, South America, Europe, Western Europe, Scandinavia, the Baltics, and Southeast Asia.

Regarding the term "America" as a cited regional name, there is general agreement about the following distribution: North America includes Canada, the United States, and Mexico; Central America includes Belize, Costa Rica, El Salvador, Guatemala, Honduras, Nicaragua, and Panama; and South America includes all countries south of Panama.

Some regional names are based on the compass. These names were logical parts of the language in early reports by travelers, explorers, and military expeditions that started mainly from Western Europe. Reports often identified areas by referring to cardinal directions noted from that part of Europe. Hence, references to regions often had terms such as "southeast" or "south-eastern." "Orient" commonly refers to areas approximately 80 degrees longitude or more eastward from the 0 degree point, the Greenwich Meridian in England. "Occident" refers to lands of the western hemisphere and Western Europe. Using such compass terms to identify all areas of the earth has been difficult, and words commonly used to indicate such locations are becoming infrequent.

A seeming puzzle of cardinal directions is that, from the United States, one normally travels westward to reach the Orient, a traditional name for China, Japan, North and South Korea, and other countries in that part of the world. For some years a U.S. airline had a rather contradictory name: Northwest Orient.[1]

In response to the repeated use by the news media of certain terms as Middle East and Southeast Asia, the U.S. Board on Geographic Names (BGN)[2] about a decade ago attempted to define these and other regions. Members studied various sources and compared systems used by scholars, libraries, and U.S. government agencies. Questions arose. North Africa might be correctly defined as the countries of Egypt, Libya, Morocco, Algeria, and Tunisia.[3] But does it include Western Sahara, the territory on Africa's western coast claimed by Morocco? Middle East could include Iraq, Iran, Israel, Lebanon, Palestine, Syria, and Turkey and the countries of the Arabian Peninsula (Bahrain, Jordan, Kuwait, Lebanon, Oman, Qatar, Saudi Arabia, the United Arab Emirates, and Yemen).[4] Far East could include China, Japan, and maybe the Koreas, but where is the Near East? For a period of time, the Near East included Algeria, Libya, Morocco, and Tunisia but now the term is rarely used. Other names with cardinal points were Southeast Asia and Southwestern Pacific.

Efforts by BGN to define the limits of such areas proved difficult. Further, practices in the United States varied. For example, the Department of State has an office called the Europe Division, whose areal responsibilities include Canada. While this was a geographic contradiction, it was accepted for reasons related to departmental management policies. Some other federal offices also categorized their activities according to nonstandard regional terms. Such definitional variations frustrated BGN efforts to define regions precisely and the project was abandoned. It was agreed, however, that such extensive regional terms could apply to a core area that perhaps centered on a specific country and then extended outward to include all or parts of contiguous countries. Thus Middle East could include the territories cited above, with a central point being Kuwait. Further, countries included in one regional name could also be part of a region having a different name.

A practical aspect of the locations of such regions arose in 1996 when lawyers near Washington, D.C., sought to identify countries that were part of the Far East. The question related to whether a restaurant whose lease required it to serve Far Eastern cuisine could also serve Vietnamese food. Research indicated that while Far East traditionally referred to China, Japan, and perhaps the Koreas, recent views say the term can also include the Philippines, Thailand, Vietnam, and perhaps Indonesia. Without any standard definition, the restaurant apparently is able to specialize in foods from various countries in that part of the world. Thus Far East indicates a larger area than was earlier the case. Presumably the restaurant could serve Vietnamese food.[5]

Another popular regional term whose areal extent may not be clearly defined is sub-Sahara. Ostensibly, the name includes all areas and countries in Africa south of the Sahara Desert, but it may be difficult to understand whether Mozambique or the Republic of South Africa is thus included.

Two other regional terms once commonly used are Mesopotamia and Levant. The former is a name given generally to the area and affected countries between and contiguous to the Tigris and Euphrates Rivers. Levant is a classical name for the coastal areas of the eastern Mediterranean, from Greece to Egypt.

REGIONAL NAMES WITHIN COUNTRIES

The United States, like other countries, has a number of regional names that lack distinct boundaries and yet are frequently used for identification purposes. In fact, the United States generally falls into four major areas: East, South, Midwest, and West. Among other smaller areas are the Northeast, the Mid-Atlantic, the Southwest, and the Northwest. No exact definitions exist for these regions. A selection of other regional names in the United States (and, in two cases, extending into Canada) include the following:

The Appalachian Mountains. A northeasterly-southwesterly mountainous feature, mainly in the United States, extending from Georgia through New England and into the Gaspé Peninsula in Canada. It consists of parallel ridges and valleys, cross valleys, and lowland areas. Another related term is *Appalachia*, which includes the middle segment of the Appalachian Mountains and is generally well-known for its rural and simple way of life.

Great Plains. An extensive, largely flat area in the United States and Canada that includes territory mostly east of the Rocky Mountains and extending north to what is called the Canadian Shield, and from the delta of the Mackensie River in Canada south to the southern part of Texas.

Midwest. An area largely between the Appalachian Mountains and the states touching the west bank of the Mississippi River and its tributary, the Red River, and north of approximately 30 degrees north latitude. The Midwest is generally understood to include the states of Ohio, Indiana, Michigan, Illinois, Wisconsin, Minnesota, Iowa, Nebraska, and perhaps Missouri and North and South Dakota. Interestingly, there is no Near West, but there is a Far West (although that term is being replaced by West). The word "Midwest" (originally Middle West) arose after settlers from areas generally east of the Appalachians increasingly moved to what became known as the West, in contrast to what was called the East. As settlers migrated farther westward, it was logical to transplant the term "West," so what was known earlier as the West became Middle West.

New England. The area composed of the states of Maine, Vermont, New Hampshire, Massachusetts, Connecticut, and Rhode Island. The name is based on the fact that many people from England settled here.

The South. Includes areas now covered by Virginia, West Virginia, Kentucky, Tennessee, North Carolina, South Carolina, Georgia, Alabama, and Mississippi. Florida is not considered part of the South.

Examples of regions whose names relate to commercial, agricultural, or cultural factors include the following:

Bible Belt. An area covering much of the South characterized by the high proportion of its population adhering to what may be called conservative Christian views.

Corn Belt. Includes sections of states in the Midwest responsible for a sizable portion of the total U.S. corn production.

Iron Belt. Area including part of New Jersey, Pennsylvania, Ohio, Indiana, and Illinois once responsible for much of the U.S. iron and steel production. (With economic changes, the Iron Belt has lost its nature as the primary source of iron, steel, and other metals.)

Rust Belt. In view of deteriorating U.S. iron and steel production, the name is now associated with the former Iron Belt.

Silicon Valley. An area south of San Francisco in California where computer chips using silicon were produced and which later became famous for its computer industry.

Outside of the United States, many regional names also exist, although they may or may not appear in any national list of names or a national gazetteer.[6] As political and cultural situations change, such names in any country may disappear, with new ones being created.

People in virtually all nations may be known by an adjectival form associated with their domestic or foreign place of residence. In the United States, a person may be a "southerner," a "New Englander," a "Mainiac," a "North Dakotan," or some other appropriate term.

Regional names, and variations of such names, are needed to describe areas, even though the territories they cover are indefinite and subject to change. Such names are a basic ingredient of communications. Combining names of regions or countries associated with ancestors to one's present country of residence for reasons of personal identification has a logical basis. Despite the value of such locational identifiers, however, some references can become less than accurate. The increasing global movement of populations makes it more difficult to develop a sense of personal geographical affiliation. Chapter 4, "How Place Names Communicate," gives more information on this topic, including how the use of double regional names can pose problems of geographic identification.

LARGE BODIES OF WATER

Regional names can also be used to identify oceans, seas, bays, gulfs, or other bodies of water that have relatively extensive dimensions and whose waters touch the coasts of two or more countries. Thus such names as the Pacific Ocean, the Atlantic Ocean, the Bering Sea, the Mediterranean Sea, the Arabian Sea, the Bay of Biscayne, and the Gulf of Mexico could be regional terms. Most such names have long histories and generally are not subject to change. As noted in Chapter 6, "Where in the World Is That Place?" however, about ten years ago during the Gulf War there was a discussion about changing the name Persian Gulf to Arabian Gulf. Furthermore, the U.S. Board on Geographic Names in 1999 voted to follow an international practice to use the name Southern Ocean for the water area surrounding Antarctica. Formerly, the board referred to that unusual and discontinuous body of water as part either of the South Pacific, the South Atlantic, or the Indian Ocean. The Southern Ocean extends from the coast of Antarctica northward to about 50 degrees south latitude.

NOTES

1. The major purpose of the airline was to connect the United States with countries in the Far East. With its principal routes beginning in the northwestern part of the United States, travel then terminated in the Orient. One could ask at what exact point does a person flying in a northwesterly direction arrive somewhere else, namely, the Orient? Perhaps this would be any point starting at 180 degrees east longitude. The airline is now called Northwest Airlines.

2. To identify such abbreviations as BGN, see Shortened Versions of Organization Names, following the introduction.

3. Algeria, Morocco, Tunisia, and sometimes Libya also may be called the Maghreb, an Arabic term defining that area.

4. In the past, Middle East may also have included areas extending from North Africa to Afghanistan, Burma, India, and Pakistan.

5. The author was consulted on this topic and, after studying the matter, gave the conclusion noted. He did not inquire about any subsequent legal action.

6. A useful source of regional names (and place names) is *Merriam Webster's Geographical Dictionary*, 3d ed. (Springfield, Mass.: Merriam-Webster, 1997).

3

Maps Say Little
without Place Names

PURPOSES OF A MAP

The word "map" has a wide variety of definitions.[1] Generally, it is a picture or graphic representation of features on all or parts of the earth's surface (or on sea bottoms or planetary surfaces). The final picture depends on the nature of surface information that is captured and on the person (a cartographer) who designs the map and selects what will be shown. Specific features on the earth include mountains, rivers, seas, forests, deserts, and other physical items. A map can also depict nations, cities, roads, bridges, and other "artificial" features associated with human activities. In addition to such static information, a map can show to the interpreter a range of dynamic elements such as population distribution, economic functions and interrelationships, natural resources, patterns of voting, climatic zones, military operations, and many other kinds of information. In these ways, maps tell about the world and its inhabitants as no other medium can. Cartographers working on undersea and planetary areas also can help scientists and others needing to know the nature of depicted features. A series of aerial photographs or satellite imagery can produce what may popularly be called a map. Although such illustrations can reveal extraordinarily useful (and often surprising) information about the nature of depicted surfaces, they most likely will need additional treatment to enable readers to interpret the captured information.

Place names may be the most important kind of detail to be added if such representations can serve intended users.[2] (The section below on U.S. mapping agencies describes a specific kind of map called a chart.)

From the earliest days of maps, place names were carried to show the locations of known and newly discovered features. In addition, some maps, albeit partially accurate but generally very imaginative, showed names of places described in various writings or legends and thought to be important. As more precise information about the earth was collected, names were needed to identify rivers, coastlines, seas, kingdoms, and settlements. Their growing use made maps a principal vehicle for place names and for disseminating information about the world and its human occupants.

Because place names identify and describe natural and humanly made features on earth, seafloor, and planetary surfaces, they significantly supplement information portrayed by maps. "Without names, a map is dumb,"[3] which metaphorically means that no matter how detailed or simple, a map cannot adequately identify features without place names.

By studying and analyzing the location and distribution of named entities, map users can also obtain a wide range of information about areas. They can understand the character of populated places and distinguish between densely populated areas and sparsely settled rural territories. They comprehend that the distribution of names can illustrate how the location of populated places is affected by the physical nature of the terrain. With such information, they can study the geographic, environmental, cultural, political, and economic relationships of areas and create programs for desired development or modification. Such knowledge can enable them to build housing developments, highways, sports stadiums, landfills, or high-tension electrical power lines in a way that conforms with human concerns. Map symbols and contours fully depict such items as settlements, administrative entities, and hydrographic and vegetational features and the interrelationships of such features, but development plans must include names to satisfy legal and other considerations.

MAP CHARACTERISTICS

A discussion of the role of a map as a major conveyor of place names justifies a brief description of the product. A primary attribute of a map is its projection, the geometric formula that gives it stipulated characteristics. Perhaps the best known is the Mercator projection, named after a Dutch cartographer who in the early part of the sixteenth century created maps of land and sea areas principally for navigational use. A major purpose of any map is to show the earth divided into defined areas by using lines of latitude and longitude spaced at designated intervals. The Mercator projection has lines of latitude parallel to the equator and lines of longitude running at right angles to lines of latitude. Lines of longitude point due north and south and are at fixed intervals from each other. As one moves from the equator to the poles, land and water areas appear much larger in north-south dimensions than they actually are because distances between lines of latitude are expanded. Also, a Mercator map covers the earth in a rectangular fashion. In other kinds of projections, lines of latitude and longitude have differing relationships and portray elements of the earth's surface in correspondingly different ways. Projections can be selected to show specific areas in designated manners. One kind of projection cuts apart the major world oceans in order to reduce distortion of land areas.

Another map element is the scale, or the relationship of a distance between two points on the map and the same two points on the ground. The scale can be expressed as a fraction. In the fraction 1/50,000, the upper figure is the distance between two points on the map and the lower figure is the distance between the same points on the ground. Translated into figures, this scale could be 1 = 1 inch and 50,000 = 50,000 inches. In other words, 1 inch on the map is the same as 50,000 inches on the ground.

Depending on the specifications of a map, names can appear in differing type styles and colors. Names printed in black capital letters normally identify nations and major administrative provinces. Upper- and lowercase black letters identify cities and other settlements. U.S. maps generally have names printed in blue letters to identify lakes, rivers, or other water bodies. Some names may be italicized to indicate a special type of feature. Thus the print style of a name can convey more than one category of information. The legend of a map defines such characteristics.

MAPPING TECHNIQUES

Traditional ground-based surveys were done using telescopes called levels to ascertain vertical heights measured from upright rods positioned at selected points. The process enabled surveyors to calculate precise elevations. Using metal lines or chains of specific lengths, they could measure exact distances between the rods. With plane tables set up over points that marked the rod positions, surveyors would draw contours between calculated vertical points on paper sheets attached to the tables, a function that would depict terrain forms. This procedure has all but disappeared for mapping large areas, but it is still employed with sophisticated ground equipment to make local surveys for property measurements.

About forty years ago, the U.S. Geological Survey (USGS) of the Department of the Interior compiled a map of part of the Western Hemisphere using high-altitude aerial photography. Although the idea of using aerial photography to make maps dated from the time when people made their first balloon flights, the modern use of this technique was heralded as the cartography of the future. The map was a highly accurate picture of the earth's surface and was thus a notable advance over traditional compilation efforts.

Aerial photography now has been largely replaced by satellite imagery of a highly sophisticated nature that can depict terrain features at larger scales and with increasing precision. Supplementing such imagery is the Global Positioning System. It is part of a portable device surveyors can carry that establishes precise ground positions in terms of latitude and longitude by interpreting data generated by orbiting satellites.

High-altitude photography and satellite imagery permit more extensive areas of the earth to be mapped with extraordinary accuracy and speed. Photography can produce maps with a spectrum of colors that reveal dramatic images of urban areas, farmlands, forests, water areas, and other kinds of surface features. Satellite imagery employs other technical measuring modes to produce data vastly different from that normally visible with the naked eye yet of substantial importance to detect mineral, soil, and hydrological conditions needed for developmental purposes. For example, the National Oceanic and Atmospheric Administration, of the U.S. Department of Commerce, has weather satellites using Advanced Very High Resolution Radiometer data to produce imagery showing ground information as forests and deserts with realistic colors.

While they can provide important and dramatic details about the appearance and nature of major (and minor) physical features, many modern maps need additional information, such as international boundaries, symbols to indicate rivers and cities, and names of countries, cities, settlements, rivers, lakes, mountains, and many other features.

Putting place names and other nonimagery information on maps requires extensive work in the field and in the office. In the United States, names are transferred from existing maps or from documents produced by state and federal agencies, but survey teams often must seek on-site information about new names or names in possible conflict. The results of such field surveys are brought to cartographic offices for filing and application to maps or charts.[4]

U.S. MAPPING AGENCIES

Maps produced by U.S. agencies are designed to meet a variety of purposes. The USGS produces topographic maps[5] that show the terrain and are considered the best vehicle to portray two broad categories of features. First, the maps depict the nature of the earth's surface by means of graphic devices called contours that show the size, shape, and elevation of terrain features. Colors (usually green) and symbols indicate vegetative cover, and blue shows rivers, streams, lakes, and other hydrological features.

Other colors and graphic patterns depict different features such as deserts, glaciers, and areas having a variety of characteristics. Second, the maps show a category of "artificial" features that includes cities and other populated places, roads, dams and reservoirs, bridges, parks, farms, and a variety of humanly made items. These are depicted by linear outlines as well as colors where appropriate.

An essential specification of topographic maps in presenting images of the world scene is to carry names of features. The precision of details shown and the number of names depend on the scale of the maps. A basic map series of the USGS is the standard topographic "quadrangle" at a scale of 1:24,000 that occupies a paper sheet approximately twenty-two by twenty-seven inches. Other series are 1:62,500 (an inch to a mile), 1:100,000, 1:250,000, and smaller scales. Clearly, the USGS, as well as other U.S. mapping agencies, must have access to current and correct names. Other agencies producing maps of part or all of the United States include the Census Bureau of the Department of Commerce, the Forest Service of the Department of Agriculture, the Great Lakes Survey of the Army Corps of Engineers, and the Bureau of Land Management of the Department of the Interior. Current place names, accurately positioned, are major requirements for all such maps.

An important type of map used for aerial or marine navigation is the "chart." With regard to aerial (or aeronautical) charts, the Federal Aviation Administration classifies two major modes of traffic that require different map styles: visual flight rules (VFR) and instrument flight rules (IFR). VFR operations rely to a large extent on a pilot's ability to identify ground features (or landmarks) from the cockpit, combined with radio contact, to assure proper flight. Charts showing terrain, settlements, natural and humanly made features, place names, and certain navigational guides are essential for planning and conducting flights. IFR traffic relies almost entirely on instruments handled by ground operators and then interpreted either automatically by flight control systems or in conjunction with pilots. Planning for flights and for in-flight operations nevertheless relies on IFR that also show details as on VFR charts but generally at smaller scales and with appreciably less ground detail.

In the early days of flight, all pilots totally depended on visual recognition of ground features to fix their courses. In the United States, commercial road maps were popular cockpit items and pilots would direct their flights along roads and use visual sightings of named places to reach destinations. In addition, names of towns with directional arrows were often painted on barn roofs for pilot guidance. Pilots flew at altitudes of perhaps several hundreds of feet, and occasionally descended to check landmarks. The U.S. agency that produces civilian aeronautical charts for the United States, its territories, and its coastal areas is the National Ocean Survey, part of the National Oceanic and Atmospheric Administration. Several private firms also produce aeronautical charts, although such charts note that they are not intended for navigational purposes.

Marine (or nautical) charts for areas in or contiguous to the United States are the responsibility of several U.S. agencies. Civilian charts for coastal waters and areas near U.S. territories are compiled and published by the National Ocean Survey. Charts for the Great Lakes and associated rivers are produced by the Great Lakes Survey. The Tennessee Valley Authority publishes charts for lakes and rivers in its area of responsibility. While enhancing safe navigation over water bodies is an essential requirement, names of harbors, islands, towns, and other land features are also important to assure recognition of associated landmarks.

The National Imagery and Mapping Agency (NIMA) of the Department of Defense produces maps for ground use and charts for air and marine navigation, principally for U.S. military purposes.[6] The agency has a prodigious production capacity that provides cartographic and related products covering the entire world.

During the Gulf War, new maps of combat areas were urgently needed to satisfy a wide spectrum of U.S. operational requirements. The maps required not only detailed information about boundaries, roads, airfields, military installations, and terrain but also names of relevant places. Cartographers at the Defense Mapping Agency (now part of NIMA) and BGN names experts worked around the clock during the crisis. While aerial operations were highly successful in eliminating the military potential of Iraq, it was the role of ground troops to advance into enemy territory. Such operations would have been handicapped without accurate reference to place names even in areas generally lacking dense populations. NIMA carried out similar functions during the political/ethnic disputes in the Republic of Bosnia and Herzegovina in 1995 and in the autonomous entity of Kosovo in Serbia in 1999.

At present, virtually all NIMA products are created automatically from satellite imagery. Place names remain a high priority. The BGN staff responsible for foreign names is housed in NIMA and its primary function is to provide current and accurate names for NIMA products. Intensive research on foreign maps and sources of names is required to obtain necessary place names. NIMA can now obtain names and cartographic information from newly independent countries that were part of the former USSR. Nevertheless, much work is required by NIMA and the BGN staff and there are still many complications. New countries have changed many names, yet maps and lists of names may not give details about names to the extent that NIMA requires for its cartographic products. In addition, BGN experts need to "romanize" names in those countries that do not use the Roman alphabet. NIMA is working with these new countries—as it does with all foreign countries where possible—to develop mutually satisfactory systems for obtaining correct names. One of the difficulties is that a number of countries do not have national agencies to standardize names.

COMMERCIAL MAPS

Commercial publishers have played an important role in producing maps of the United States and other areas. Early maps covered territories being explored as well as existing cities (many with a bird's-eye perspective). At present such commercial products include large-scale topographic maps extracted from public sources and modified with additional information to meet, for example, tourist needs. Private publishers have been responsible for reaching the public and educational institutions with important products. The Internet also provides access to maps with various contents. Perhaps the best-known commercial map publisher is Rand McNally, which started business in 1856. The National Geographic Society also is a major producer of maps, initially as a nonprofit body. Recently its cartographic operations were reorganized as a commercial enterprise. Such organizations and other publishers of maps, globes, and atlases depend heavily on accurate place names. One of world's most comprehensive atlases is the British *Times Atlas of the World*,[7] whose 210,000 place names exceeds that of any other similar publication. During the last decade, private mapping organizations in the United States have depended heavily on BGN for current names to revise their products. With the collapse of the Soviet Union and the manifold changes of names, some publishers recognized that their new atlases could soon be out of date and offered to send purchasers free pages with new names and boundaries.

ECOLOGY AND MAPS

One activity of increasing interest today is identifying the habitat of wildlife and tracing its migrational paths. Any report of a newly discovered creature or a creature far from its normal habitat normally will cite the location of observation. While this can be done with latitude and longitude, a report should give the location in terms of its distance and direction from the nearest place, probably a town or a village. Depending on how well a named place is known, people may have to consult maps or other sources to fix the location. Names experts at NIMA once received an inquiry from the staff at the Smithsonian Institution asking about the location of a place mentioned by an ornithologist in a report about a bird he had observed in Central America. The bird was far removed from its normal habitat or migration path. The names experts at NIMA found the cited place in its extensive files of names and informed people at the

Smithsonian Institution of its map location. By knowing the location, ornithologists could then carry out additional studies that might indicate a new or expanded habitat or migration route for the species.

NOTES

1. Persons with computers can open a glossary of cartographic terms that include over 300 definitions of the word "map." Some of the definitions are from sources dating back 300 or more years. To locate this information, enter <http://geography.miningco.com/msub39.htm>

2. Early in the year 2000, a satellite project called Endeavor produced photographic images of much of the earth's surface with details previously not possible. The project was a joint effort of the National Aeronautical and Space Agency and NIMA and will permit scientists and others to analyze and interpret a host of environmental factors related to the earth and its inhabitants. Although the images are called maps, they require additional information such as place names to be fully useful.

3. Mentioned by Francisco Gall, the leading names authority in Guatemala, who worked with the author on international names programs from 1973 until Gall's death in 1987.

4. The U.S. Board on Geographic Names is responsible for coordinating all such names to assure that they meet national standards. In chapter 8 the board is described in detail.

5. The topographic maps are produced by the National Mapping Division of the USGS in coordination with related state programs.

6. NIMA was formed in 1996 by merging the Defense Mapping Agency with elements of other U.S. government agencies producing satellite imagery to meet intelligence requirements. Since the offices were making similar if not identical products to varying degrees, it was deemed logical and expedient to combine such operations.

7. *Times Atlas of the World*, comprehensive ed. (London: Times Books, 1992).

II

How Place Names Affect Us

The next three chapters describe the ways in which place names have become an important part of human life. Names do much more than inform us about the physical nature of our world. They can also tell us about the human aspects of places, and they have become significant parts of our daily vocabulary. Further, place names have been recognized for the significant connections they have made to the lives of many people, including farmers, commercial entrepreneurs, and scientists. The impact on authors, musicians, painters, and photographers is especially notable as demonstrated by the way they have used place names to associate their creations with world areas, large and small.

4

How Place Names Communicate

INFORM US ABOUT OUR WORLD

As indicated in the preceding chapter, a major purpose of a place name is to provide information about natural and humanly made features on the surface of the earth and natural features on the seafloor and planetary surfaces. The kinds of information vary considerably.

In certain cultures, place names may describe a human or natural activity. A review of a book about the Western Apache provides interesting background.[1] Two place names of the Western Apache are Water Flows Down on a Succession of Flat Rocks and Lizards Dart Away in Front. The function of such names can be to cite experiences of ancestors. As a non-writing society, the Western Apache depended on place names to provide an oral history, which otherwise would be lacking in useful details. Such naming practices are common to Native Americans as well as indigenous people in other parts of the world.[2]

IDENTIFY ASPECTS OF OUR LIVES

Place names are essential to many aspects of our individual and collective lives. When people apply for credit, a driver's license, or anything that requires personal identification, they must give the address of their residence, which includes a town and its state. Some registration forms also require one's place of birth. This information thus becomes part of our life history and provides adequate proof of our individual existence.

People meeting for the first time will ask two questions early in any conversation: What is your name? and Where are you from? The answer to the second question can provide essential personal background information. Furthermore, we all have probably experienced a noticeable reaction when we learn that a stranger is from our own hometown. Even though in today's mobile society it is increasingly likely that one person can call various places "home," it is always interesting to note which place a person may identify. Most take pride in their birthplace and may relate more quickly to others from the same town, as opposed to those from other, more distant areas.

Some places may be associated with histories that have changed so radically that people who lived there may not want to admit it or may have to explain that certain factors have been altered. A woman once told the author that she sometimes encountered negative reactions from people in her synagogue when she mentioned her place of birth. She was born in Palestine, under the British mandate from 1920 until 1948, at which date the territory was divided between Jordan and Israel. Given some of the current Israeli views about Palestine, the responses she noted were sometimes less than positive. The author recommended she should say she was actually born in that part of Palestine that is now Israel.

Genealogical research also relies heavily on place names as evidence of one's ancestry and their movements with time. People long removed from places where their parents, grandparents, or other ancestors were born may attempt to visit the towns or villages where they lived. A visit to such areas can awaken emotions that strengthen their sense of belonging to a place. Individuals who cannot easily identify their place of birth often try to establish ties with their roots. Even seeing a place name on a map or in a personal letter can communicate sentiments. The name of one's birthplace can evoke pleasant associations. Some names can also become a symbol. Psalm 137 in the Bible tells of people from Jerusalem who were forcibly moved to Babylon. Despite their captivity and the destruction of their city, the people vowed never to forget Jerusalem. This strong tie to the name of a place remained constant over the centuries.

Other place names can generate very unpleasant memories. Examples are Buchenwald and Auschwitz (the former German name for Oświęcim, the current Polish name), two of several German concentration camps during World War II.

National place names have the power to communicate at the personal level. When used as adjectives, such names can convey an association with a number of factors such as political and economic situations, religious backgrounds, ethnic characteristics, historical events, and even climatic and terrain conditions. People commonly identify others by their nationality. She is "French," he is "Italian," they are "Norwegian," the musician is "Russian." These are not, strictly speaking, place names. But by identifying homelands they give a tie to a place deemed important.

For many years, people wanting to immigrate to the United States did so because they wanted to become "Americans." U.S. citizens identified themselves as "Americans" with a sincere sense of patriotic pride. People in other nations do the same. A hundred years ago, the Gilbert and Sullivan operetta *HMS Pinafore* called attention to the concept of national pride when the chorus lauded the honorable status of a particular seaman by singing "He is an Englishman."

Many U.S. citizens also take pride in their background as Irish-Americans, Italian-Americans, German-Americans, Welsh-Americans, or other terms with dual nationality connotations. Some groups in the United States tracing their ancestral origins to other areas also have developed hyphenated terms to identify themselves. Typical terms combine two regional names, as African-American (or Afro-American), Asian-American, or Pacific-American.[3] Native American is a term commonly applied to indigenous people whose ancestors, before European settlement, lived in what is now the United States. A new term is Euro-American to describe those with European ancestors.

These adjectives may have near official status for official registration purposes and have become part of the current vocabulary in many contexts. The practice is in response to a need a growing number of people have to identify their roots. Children registering for school may have to identify their parents in such terms to assure a proper minority ratio. But many believe the concept raises complications. Individuals with parents and grandparents from different backgrounds logically could add two or more additional regional (or national) terms to the word "American-." It is possible a child could be classified as an Afro-Latino-Asian-Indigenous American. This procedure could cause much confusion as to a person's cultural or ethnic background.

The meaning of certain national terms may raise questions. Using the word "American" to identify one's national affiliation (particularly in the United States) may imply that America had been known by that name from its first settlement in prehistoric times. Its use is, therefore, subject to question by certain groups who feel that concept should be changed for the sake of anthropological accuracy. But the application of the name "America" to the entire Western Hemisphere—North America, Central America, and South America—makes its use virtually permanent because it effectively communicates messages as to location, if not culture. On the other hand, only citizens of the United States may use the term "American" to identify themselves. People elsewhere in the Western Hemisphere may refer to themselves by national (or provincial) terms.

Further, the term "Indian" has come under criticism. Its use originated when Columbus wrote of people he encountered as Indians because he believed he had landed on or near the shore of India. Some believe the term "Indian" should be abandoned in favor of others such as Native American or Amerindian. But those terms have problems. Logic could say that anyone born in the United States is a Native American. Amerindian is seen as a composite of two unsuitable adjectives. Increasingly, those whose ancestry predates European settlement are referred to as "indigenous peoples." On the other hand, there appears to be a growing sense among "Native Americans" that the term "Indian" actually is acceptable. The existence of a Bureau of Indian Affairs in the Department of the Interior attests to that view.

The use of hyphenated regional names can raise questions of geographic if not ethnic inconsistencies. The expression "African-American" was introduced so people previously defined as black, colored, or Negro could remove what many saw as negative connotations. The motive is understandable but the term "African"[4] can legitimately apply to anyone in, or coming from, the entire continent of Africa, which includes a variety of ethnic groups, including Arabic, Bedouin, Berber, Chinese, Dutch, English, and Hindu. The current usage of the term "African-American" (or its variant, Afro-American) appears to relate only to those whose ancestors came from Africa south of the Sahara under conditions of slavery.

Terms such as African, American, English, Italian, and other adjectival forms of country or areal names are important to many kinds of communications, as are regional names such as Southeast Asia and South America. Although it is a commonly accepted goal that all geographic names should relate to specific and well-defined places, it is not always possible. As indicated above, there are differing understandings of such terms as African. News accounts commonly refer to regional place names, yet unless they mention the countries involved, it is unlikely that the public will clearly understand where the named areas are. The idea that an adjectival geographic name can identify the background of a person is subject to some doubt. In England many Pakistani immigrants live in the city of Bradford. Is anyone of this population English? In Germany, children of immigrants from Turkey may retain Turkish cultural habits, but are they not Germans? The increasingly mobile global population complicates the notion that one's place of birth can adequately identify one's culture on a permanent basis.

Another question relates to citizens of Israel. It is generally believed that the term "Israeli" refers to citizens of that country who are Jewish or to a national point of view that reflects a Jewish perspective. Thus the statements "he is an Israeli" and "the Israeli position" generally have clear connotations. In fact, however, Arab citizens of that country also are Israelis.

Many U.S. citizens do not approve of using hyphenated terms to describe their ancestry. They believe that once citizenship has been granted, individuals and their children born in the country are ipso facto Americans. They also believe that their ethnic heritage is a personal matter. The author once heard of a case relative to this issue. A person of Hawaiian background employed by a U.S. agency was asked to be interviewed for an article in an internal magazine publicizing "Pan Asians." She refused, saying she was an "American."

PROVIDE HISTORICAL INFORMATION

Place names can impart historical and cultural information about both small and large areas. A study of place names in virtually any country can reveal historical patterns of settlement by various ethnic groups. For example, in Brittany (in France) and in Wales (in the United Kingdom) names of towns reflect the distribution of Celtic people in those areas hundreds of years ago. A French topographic map at the scale of 1:25,000 shows that the area in Brittany around Quimperle has dozens of place names beginning with C, G, K, L, and Q. Clohars-Carnot, Guerveur, Kerganaouen, Kergloarec, Kervaveon, Lisloc'h, and Quililoen are examples. Such names are of Celtic origin and are relatively rare in other parts of France.

A U.K. topographic map at 1:50,000 that covers part of Wales shows such towns as Caernarfon, Caergeiliog, and Carreglefn. The relationship to a common linguistic or cultural past in both areas is evident. England also has names reflecting invasions and settlement by Angles, Danes, Jutes, Romans, Saxons, and Vikings. In many cases, names have undergone considerable phonetic modifications. The origins of names in many world areas reflect similar historical influences.

Many place names in the United States communicate interesting historical facts. The book *Albion's Seed*[5] notes that immigrants coming to the eastern part of America from four areas of England during the seventeenth and eighteenth centuries brought distinctive habits of speech, religion, and many other customs that had lasting effects on the new country. The immigrants applied place names that, while all being of English extraction, nevertheless in many cases reflected the differing attitudes in the areas in England where the newcomers formerly lived.

Those coming from eastern England and settling in the Massachusetts Bay Colony generally used names of towns or settlements from that part of England. Included are Braintree, Cambridge, Dedham, Groton, Haverill, Ipswich, and Newbury. Settlers from southwestern England moving into Virginia and the tidewater areas applied several names reflecting their attitudes, which included a view favorable of the English aristocracy. In 1703, counties in Virginia were King and Queen, King William, Elizabeth City, Henrico, Prince George, Charles City, James City, and Princess Anne. Other names were for features or places in that part of England, including Gloucester, Isle of Wight, Middlesex, Northampton, Stafford, Surrey, and

Warwick. English immigrants (largely Quakers) coming from the upper central part of England and settling on the banks of the Delaware River selected names for new towns reflecting their social concepts. Some names were Philadelphia, Salem, Concord, Upper Providence, and Nether Providence. County names applied in Pennsylvania, northern Delaware, and New Jersey also had names used in parts of England, including Burlington, Chester, Lancaster, Monmouth, Newcastle, and York.

Immigrants from the fourth area of England—the hilly region between England and Scotland—generally settled in Appalachia. They did not import names honoring any aristocracy or having elite connotations but instead chose a variety of names reflecting English places. Cumberland, the name of the northwestern-most county in England, is found as a name or part of a name in Kentucky, Maryland, North Carolina, and Tennessee. A selection of other names in the cited fourth area that were applied to Appalachia and associated areas includes Aberdeen, Abernethy, Ayr, Balfour, Belfast, Caledonia, Coleraine, Donegal, Durham, Galloway, Londonderry, Newcastle, New Dublin, New Glasgow, and New Scotland. The rural and somewhat isolationist nature of settlers in Appalachia also generated numerous colorful names, including Clabber Branch, Logan's Station, Pinchgut, and Worry. In addition, clan names were numerous. Such nomenclature reflects the nature of the peoples' existence in their earlier homelands.

Later immigrations of Dutch, French, and German settlers also affected the names of places. A look at maps covering Pennsylvania, Delaware, and New York show names that indicate such origins. The Schuylkill River in Pennsylvania is related to the presence of Dutch settlers in the eighteenth century. The original name was Schuyl Kill. Schuyl was probably the name of a settler, and Kill meant river or stream in Dutch. Since the English language predominated, the two terms were joined as Schuylkill, and then River was added.[6] In other areas, such as California, Florida, and New Mexico, Spanish explorers and settlers implanted a grid of names that prevail to this day. The book *Exploring the Beloved Country*[7] by Wilbur Zelinsky includes a number of articles that describe the relationship between place names and human settlement in America. Virtually all countries have names that reflect patterns of migration, and a number of books deal with this topic. One major work is *Names on the Land* by George R. Stewart.[8] Thus place names can communicate information about the history and culture of virtually any area of the earth.

DESCRIBE THE NATURE OF TERRAIN

The distribution and nature of names convey information about the character of the terrain and even land-use practices. Furthermore, words on maps that are not place names but describe feature types also communicate. For example, words such as "swamp," "farm," "estate," and "forest" may be applied to identify natural and humanly made features. Such words can be considered "generic terms" that describe feature types. In fact, all place names could consist of nothing more than such terms, but in a world of maps and extensive communications, people generally seek to add a name to a generic term for more precise and useful identification. Nevertheless, some cultures may refer to locally known features only as "the river" or "the mountain." Even in the United States, individuals may feel no need to refer to a nearby feature in specific terms. Chapter 11 discusses this topic in more detail.

COMMON WORDS

Many place names have become words in our daily vocabulary. Books on the origin of words include this subject and illustrate how the English language has evolved and continues to do so.[9] Dictionaries also may identify words that are derived from place names. Some words trace their origins to perhaps ancient times and have been modified through linguistic processes. Although few people now think about such roots and make no association between words and places, many such words—used mainly as nouns but also as adjectives and even as verbs—are part of our vocabulary. They might be called "name-words." A number of fabrics, food, and beverages got their names either because they originated at a specific place or because of a logical geographic association. For cheeses and wines, France may be the most common source of names.

The following contrived statement shows how such words could be used as nouns, adjectives, or verbs in normal conversation. The italicized words are derived from place names. "Wearing *cordovan* shoes, *argyle* socks, a *cardigan* sweater, a *tweed* jacket, a *cravat*, and a *homburg*, a man rubbed eau de *cologne* on his face and then drove in his *limousine* to attend a party at a neighbor's home. There he *meandered* around and saw other guests wearing *tuxedos*, *cashmere* coats, *denims*, blue *jeans*, *Mackinaw* jackets, and *derbies*. Then he noted that the availability of beverages included *scotch*, *bourbon*, *burgundy*, *champagne*, *port* wine, and *sherry*, and an assortment of cheeses as *camembert*, *roquefort*, *swiss*, and *brie*. *Cantaloupe* was served before dinner. The main course featured eggplant *parmigiana*, *brussels* sprouts, *wieners*, and *hamburgers*, all served on elegant *china* arranged on a *damask* tablecloth. *Tabasco* sauce was available. After a dessert, he enjoyed a cup of *mocha*."[10]

Appendix A gives the background of the cited words and many more. It does not include all possible name-words or brand-names per se. The number of names for cheeses and wines readily indicates how people identified locally made products of these kinds with their places of origin.

ANIMAL NAMES

Place names are parts of breed names of many animals. Angela Ray wrote an article in *Names* saying that of 136 breeds of dogs listed by the American Kennel Club, 86 have a geographic name or its variant identifying the area where they were first bred or found. Examples are Afghan, Alaskan Malamute, American Foxhound, Boston Terrier, Cardigan Welsh Terrier, Chihuahua, Dalmatian, English Setter, German Shepherd, Great Dane, Irish Setter, Labrador Retriever, Maltese, Pekinese, Pomeranian, Rottweiler, Shetland Sheep Dog, Siberian Husky, Weimaraner, and Yorkshire Terrier.[11]

Some horses are identified by place names. Perhaps the most famous is the Arabian, native to the Arabian Peninsula and long valued for slender beauty and speed. Another is the Lipizzaner, a breed used by an Austrian group for incomparable show performances. Its name is from Lipizza, a place formerly part of Italy but now in Slovenia (the name is Lipica), where the horses were first trained. The Tennessee Walking Horse is associated with that state. The Percheron is a breed of horse from La Perche, France. It has a stalwart frame and shaggy mane and forelocks and is noted for its ability to pull heavy wagon loads.

Many cows have names based on their original breeding places. The most famous milk cows are Guernsey and Jersey (both from the British islands with those names in the southern part of the English Channel near the west coast of Brittany, France), and Holstein, bred originally in Friesland (part of Holland) but also associated with Holstein, part of Germany at the base of the Jutland Peninsula.

FAMILY NAMES

Early naming habits probably gave only one name to a person, which served mainly as a "first" name. When populations increased and more precise identification of individuals was required, children in some areas might be given a "last" name. The pattern of personal naming has, in any case, a complex history. And even today practices vary. In the past, when a woman in the United States married, she took her husband's last name, but today the wife may keep her own name, or both mates may take a third and different name. The book *The Language of Names* provides extensive information about personal names.[12] A number of "last" names, also known as family names, may have origins associated with places. In Biblical times, the name of a person's place of birth was often mentioned for better identification. Jesus of Nazareth and Saul of Tarsus are principal examples. Other last names reflected places or types of features associated with the residence of a family.

The book *The Mother Tongue: English and How It Got That Way*[13] notes that when the custom of adopting last names got under way in England, four major sources were place names (as Lincoln and Worthington), nicknames (descriptive of a person, as Whitehead or Armstrong), trade names (as Farmer and Carpenter), and patronymic (father's last name, as Johnson and Peterson). A casual review of a telephone directory in England, Canada, and the United States will show many last names that are the same as those of natural and humanly made features. Without studying individual names in detail, it would be difficult to determine which places are named for people and which family names are derived from named places. Also common are family names derived from generic terms describing such terrain features as bank, beach, cliff, field, forest, hill, lake, plain, and river.

In England, the system of titles includes the name of a place. The person inheriting or being bequeathed the right of domain over a territory will have a title that includes the name of the area. For examples, the Earl of Derby exercises domain over a place called Derby, and the Duke of York has domain over a territory named York. While persons so privileged also have a standard name prior to being entitled, the formal name more or less supplants the former. Thus Edward Stanley, 1752–1834, was the twelfth Earl of Derby. The territories can be vast, can include towns, districts, and counties, and may be enlarged by purchase or other arrangements.

AUTOMOBILE NAMES

Automobile makers, particularly American ones, for a number of years have produced cars often better known by their model names than by their standard names. One dominant idea has been to add a nickname that would evoke the image of a popular resort or a place well-known for other reasons (e.g., a race course and gambling casinos at Monte Carlo give that name a certain appeal) and thus upgrade the desirability of the vehicle.[14] Such names applied to passenger cars (and increasingly to SUVs) have included Buick *Park Avenue* and *Riviera;* Cadillac *Burgundy, De Ville, Eldorado,* and *Seville;* Chevrolet *Corsica, Malibu, Monte Carlo,* and *Tahoe;* Chrysler *Concorde, Cordoba, Landau, New Yorker,* and *Sebring;* Dodge *Dakota, Daytona, Durango,* and *Monaco;* Ford *Grand Torino* and *Contour* (not a place name but a term with a geographic implication); GMC *Savana* (possibly a modified spelling of Savannah, a river and a place in Georgia), *Sierra* (Spanish word for mountain range, but also part of place names in the United States and Latin America), *Sonoma, Yukon,* and a new 1999 SUV named *Yukon Denali;*[15] Jeep Wrangler *Sahara* and Cherokee *Laredo;* Lincoln *Continental;* Mercury *Capri;* Oldsmobile *Calais* and *Firenza;* and Pontiac *Bonneville.*

The Japanese car maker Toyota has adopted the practice and has models named *Avalon, Tacoma,* and *Tundra* (a name given subarctic regions with cold climates). The Isuzu brand has a model called *Oasis* (a word meaning a desirable spot with water in a desert). Otherwise, Japanese companies rely heavily on nonword names such as *Corolla, Celica,* and *Tercel.* The German Volkswagen had a model called *Scirocco* (the name of a wind associated with Italy). Another model is *Golf,* which when translated to English is either golf (the sport) or gulf; in both cases the name conveys a certain aura of desirability. The South Korean Hyundai has a model *Tiburon,* a place in California near San Francisco.

The custom of using place names for automobiles may be dying out. Car names increasingly are either words conveying an attractive message (such as *Geo, Probe, Festiva, Escort, Omega, Cougar, Sable,* and *Tracer*) or contrived words that have no specific meaning but seem relevant (such as *Lumina*). The April 2000 issue of *Consumer Reports* evaluated 205 vehicles, of which about thirty-five had car nicknames or names that relate to a geographic feature in a positive way. This number is slightly less than that in the April 1999 issue.

FOOD AND BEVERAGE NAMES

The names of foods and beverages also rely on place names, real or fictional. Florida oranges are publicized as such because of claims they are superior to oranges from other places. A well-known mineral water is Evian, the name of a resort in France on Lake Geneva. Another water is Saratoga, produced in Saratoga Springs, New York, that has long been a resort. Most mineral waters have names that evoke images of natural areas untainted by human settlement, and labels that often show mountains and rushing streams. Some, in fact, have such names and labels even though the contents are artificially produced.

RAILROAD NAMES

Railroad companies almost universally used place names as their corporate titles. A source of railroad names[16] shows that of approximately 7,300 railroads operating over the past 150 years (many no longer exist), about 95 percent had one or more place names in their official corporate names. The first name in the list is Abbeville & Waycross, carrying names of towns in Alabama and Georgia, and the last name is Zwolle & Eastern, with a name of a place in Louisiana and a cardinal direction indicator. The main purpose in selecting place names was to associate a railroad with the area it served. Another purpose, perhaps, was to establish the reputation of the railroad by using the names of places, some locally familiar and some known regionally or nationally.

Names of states, towns, counties, lakes, rivers, valleys, and seas were common and modifiers such as central, northern, southern, eastern, and western were sometimes added, possibly to expand the geographical association. Some well-known names are: Atchison, Topeka, and Santa Fe; Baltimore and Ohio; Chesapeake and Ohio; New York Central; Northern Pacific; Norfolk and Western; and Union Pacific. Country songs also used actual and fictional names of railroads to generate a nostalgic feeling not only for a mode of transportation but also for a lifestyle in certain areas. Two well-known songs are "The Chattanooga Choo-Choo" (actually, not an official railroad name) and "The Atchison, Topeka, and the Santa Fe."

Railroads often adopted nicknames for convenient identification. Some nicknames included a reference to regions they traversed. The Chicago, Indianapolis & Louisville railroad had the nickname the Hoosier Line. (While railroads selected place names for their corporate titles, many communities owe their names to nearby lines. As railroads ran in sparsely settled areas, towns serving the lines were born. In the process, railroad employees often named them.)

Railroad names are less known today because the network of railroads has diminished. In 1976, the federal government created Conrail as a private concern to carry all freight. Conrail trains may, however, include individual cars with names associated with formerly individual companies. A few years earlier, the federal government formed Amtrak as the nation's single passenger railroad. Except for lines it owns in the eastern part of the country, it operates on tracks still owned and maintained by private railroads. In addition to Conrail and Amtrak, there are a number of local private lines that operate for tourism. Although railroads no longer play the role in American culture they once did, the use of names of states, towns, and natural features for railroads clearly demonstrates another association that place names have with human enterprises. Despite consolidation of many rail lines, there is recent evidence that some companies are getting back into business with their original names.

MINERAL NAMES

There are some 1,500 kinds of minerals, of which perhaps only about 200 are significant. In any case, names of 334 are based on place names, reflecting the fact they were discovered or identified in the indicated areas. In keeping with the requirement of having "ite" as the suffix for mineral names, the spelling of place names may be modified before adding the suffix, for example, Amazonite (Amazon River, Brazil), Canadite (Canada), Ceylonite (Ceylon, now Sri Lanka), Germanite (Germany), Ozarkite

(Ozark Mountains, United States), Tasmanite (Tasmania, Australia), Uralite (Ural Mountains, Russia), Utahlite (Utah, United States), and Valencianite (Valencia, Spain). The names are English forms mainly because that language was the principal tongue in the scientific world when most of the minerals were discovered and named.[17]

NOTES

1. Douglas Duer, review of Keith Basso, *Wisdom Sits in Places: Language and Landscape among the Western Apache* (Albuquerque: University of New Mexico Press, 1996), in *Professional Geographer* 50, no. 1 (1998): 254–255.

2. Such long names are in some cases also part of current names in the United States. For example, Los Angeles is taken from the earlier name given by Spanish settlers: El Pueblo de la Reína de los Angeles de la Porciúncula.

3. Pan-Asian is another term. It is supposed to identify American citizens whose original habitat was somewhere in the Pacific region, and probably is no different than Pacific-American.

4. *The American Heritage Dictionary* defines African-American: "An American of African ancestry; an Afro-American." The book also has "usage notes" on the word "black" that discuss the variations in use and meaning of such terms.

5. David Hackett Fischer, *Albion's Seed: Four British Folkways in America* (New York: Oxford University Press, 1989).

6. This name is an example of what might be called a "redundant" name because it contains two words having the same meaning: kill and river. English settlers and travelers normally made a single term out of existing names in other languages that had two or more words, and then added a separate generic term. Other examples are the name Rio Grande River, which essentially means "River Large River," and the Hwang Ho River in China, which means "River Hwang River." There are many other such cases.

7. Wilbur Zelinsky, *Exploring the Beloved Country: Geographic Forays into American Society and Culture* (Iowa City: University of Iowa Press, 1994) includes several segments on the nature of place names in America from a range of interesting perspectives. For example, one part, "Classical Town Names in the United States," describes the reasons many settlements received such names as Athens, Troy, Corinth, or Ithaca. His writings over the past several decades on the geographic elements of American culture make him an outstanding expert in this area.

8. George R. Stewart, *Names on the Land* (1945; reprint, New York: Random House, 1972).

9. Basic sources are Joseph T. Shipley, *Dictionary of Word Origins* 2d ed. (New York: Philosophical Library, 1945); and Wilfred Funk, *Word Origins and Their Romantic Stories* (New York: Bell, 1978).

10. Mocha is a kind of coffee named for a port in the Yemen Arab Republic. Java, a kind of coffee from Indonesia, is also a common name for coffee in general.

11. Angela G. Ray, "Calling the Dog: The Sources of AKC Breed Names," *Names* 43, no. 1 (1995): 3–28.

12. Justin Kaplan and Anne Bernays, *The Language of Names* (New York: Simon & Schuster, 1997).

13. Bill Bryson, *The Mother Tongue: English and How It Got That Way* (New York: William Morrow, 1990).

14. Ingrid Pillar, *American Automobile Names* (Essen, Germany: Verlag Die Blaue Eule, 1996), 148–149, 176–183. The author notes that Plymouth was chosen as a principal car name because it related to the place where the Pilgrims landed and therefore evoked a connection with a determined people who conquered difficult circumstances.

15. Adding the name *Denali* to an SUV earlier called *Yukon* is intended to produce an impression of an attractive yet adventuresome place. But the two names pose a geographic puzzle. Denali is a name favored by many as the historically correct name of Mount McKinley, which is the official name of the mountain in Alaska. Yukon is the name of a Canadian territory contiguous with Alaska but located a considerable distance away from Denali. Yukon is also the name of a river flowing from that territory into Alaska but is not closely associated with Denali. Yukon can thus refer to two features. It is not likely many owners of the SUV will question whether the names apply to a single geographic entity.

16. *Railroad Names*, 3d ed., compiled and published by William D. Edson, Potomac, Md.: McClain, 1993.

17. Breandan S. MacAodha, "Mineral Names from Toponyms," *Names* 37, no. 1 (1989): 19–30.

5

Place Names and the Arts

LITERATURE

Marcel Proust provides a unique description of the power of place names in his massive *Remembrance of Things Past*. His book includes numerous references to names of cities, towns, railway stations, rivers, and other features he visited, often on a regular basis, and he describes in detail how they formed lasting impressions. One key statement regarding places is, "But I need only, to make them reappear, pronounce the names."[1] Another comment, written in the complicated manner typical of Proust but simplified here, is as follows: "it is not only to towns and rivers that names give an individuality. . . . it is not only the physical universe which they pattern with differences. . . . there is the social universe also."[2] It is unlikely that Proust studied place names in a scholarly manner, yet by associating a range of physical, psychological, and personal characteristics with named places, he reflects the views of experts in the field.

The titles of many books also refer to places. In *Dakota: A Spiritual Geography*,[3] Kathleen Norris tells about her decision to return to her family's home on the border of North Dakota and South Dakota. She and her husband agreed to move there from New York City in the 1970s after having gone there earlier to settle the estate after the death of her grandparents. She was captured by the way of life in that region. Her description of the area includes not only its geographic attributes—the climate, the terrain, the rural atmosphere, and the scenery—but also its human characteristics.

The combination was radically different from her previous experiences in the East and, although not all factors were pleasant, she and her husband resettled there. Without question, the book provides a realistic description of a place identified as the Dakotas and an associated way of life.

Lake Wobegon Days, by Garrison Keillor, recounts tales about the lives of people in or near a town in Minnesota he called Lake Wobegon.[4] His book describes the settlement in the kind of detail that might be found in an official guidebook. Chapter 20 of the present text, "Unusual and Unacceptable Names," gives a more thorough picture of the place.

Poetry

In English and American literature, especially in the last 150 years or so, poets have written about many places and features associated with personal or historical events taking place there. This section mentions only some examples. Stephen Vincent Benét's poem "American Names" is most appealing because it recites his fascination with the array of unusual and colorful place names in the United States. The first verse is:

> I have fallen in love with American names,
> The sharp names that never get fat,
> The snakeskin titles of mining claims,
> The plumed war-bonnets of Medicine Hat,
> Tucson, and Deadwood, and Lost-Mule-Flat.

The last line of the poem expresses his wish to be interred at a place having an equally appealing name: "Bury my heart at Wounded Knee."[5]

Henry Wadsworth Longfellow's poem "Hiawatha" begins with "By the Shores of Kichigoomee," a lake bordering Minnesota now called Lake Superior. Nathaniel Hawthorne wrote of the forced exodus of people from a region in Canada then known as Acadia. (People resettling in Louisiana became known as "Cajuns," a derivation of Acadia, the name of the Canadian region in the Micmac Indian language.) Sidney Lanier wrote "Song of the Chattahoochee," a river in Georgia. In this poem, he describes in detail the characteristics of the river and of features associated with places called Habersham and Hall that relate to parts of his life. Carl Sandburg's poem "Grass" refers to the war dead at Austerlitz, Waterloo, Gettysburg, Ypres, and Verdun. Vachel Lindsay wrote "Congo," a poem about life in Africa as he understood it. He annotated the work with voice or sound accompaniment. The short poem "Night Song at Amalfi" by Sara Teasdale includes an indirect reference to a body of water near that town in Italy. The poem "London Nightfall" by John Gould Fletcher is a short description of that city as night approached. William Butler Yeats composed the poem

"The Lake Isle of Innisfree" to identify a place in Ireland whose surroundings attracted him to live there. "Rounding the Horn" by John Masefield describes the trials of a ship crew passing Cape Horn during a storm. James Russell Lowell wrote the poem "Ode to France," which narrates the development of that nation. He mentions Paris, the Pyrenees, and the Rhine. "Appledore" is his poem about a coastal place in New England with its rocks, beaches, and storms.

Perhaps one of the most famous poems based on a place is "In Flanders Field," a work by John McCrae honoring the fallen soldiers of World War I in an area that is part of Belgium and France. G. K. Chesterton wrote "Lepanto," a work about a major sea battle in 1517 at a place known by that name in Greece.

John Greenleaf Whittier composed poetry referencing many places in America and elsewhere as he wrote about issues and people of the times. Included as titles, verse elements, or both, are Atlanta, California, Chicago, Delaware, Lake Superior, Maine, Maryland (and Frederick), Massachusetts (and Barnstable, Boston, Hampton River, the Merrimack River, Mount Monadnock, Penntucket, Nantucket, Newbury, Wachuset, and Worcester), Oregon, Pennsylvania (and Gettysburg), the Potomac River, the Sierra Mountains, the Red River, Texas, Virginia (and Roanoke and Yorktown), and Washington, D.C. Places outside the United States include Aberdeen, Brazil, Brussels, Ceylon, Cuba, England, Finland, the Holy Land, Labrador, Lebanon, Naples, Newfoundland, Perugia, Rome, and St. Georges Bank. Whittier's reliance on names was essential to set the tone and the relevance of his poetry to places. He often described the natural settings of places in great detail.

Plays

Numerous plays refer to places in both title and text. Those written by Shakespeare provide good examples. He frequently used place names, ranging from names of countries to those of local villages or natural features and even some legendary places. The reasons for using names are clear: it was necessary to identify settings for the dramas; the plots required that characters relate to a country, a province, or a town or village; specific scenes occurred at designated places; references to far-off places were necessary to develop an atmosphere relevant to the play; and mythical places or areas known in ancient literature were referred to in order to develop an analogy or a theme.

A random selection of only a few of Shakespeare's works illustrates his use of names. In act 3, scene 6 of *Antony and Cleopatra*, Caesar mentions Rome, Alexandria, Egypt, Syria, Cyprus, Lydia, Media, Parthia, Armenia, Cilicia, Phoenicia, Sicily, Libya, Cappadocia, Paphlagonia, Thrace, Arabia,

Pont, Comagene, Mede, and Lycaonia. In the next scene, Antony refers to Tarentum, Brundusium, the Ionian Sea, Toryne, and Pharsalia. A 1997 performance of the play in Washington, D.C., opened an act with a large map showing place names of the Roman Empire that Caesar pointed to as he spoke his lines. The play *Macbeth* refers to places close to the locale of the play or connected to it: Birnam Wood (a forest), Cumberland, Fife, Ireland, Norway, Scotland, the Western Isles (islands to the west of Scotland and perhaps Ireland), Scone, and Scottish areas (as Cawdor, Glamis, and Ross). The play also mentions Allepo, Arabia, and Golgotha. In *Much Ado about Nothing*, the first line refers to Aragon and Messina, which quickly establishes Italy as the location of the play. Additional places that are mentioned are Padua and Italy. In *The Merchant of Venice*, references are made to Ballario, Barbary, Belmont, England, India, the Indies, Lisbon, Mexico, Padua, the Rialto, and Tripolis. In *Hamlet*, places include Barbary, Denmark, Elsinore, England, France, Israel, Normandy, Norway, Poland, Rome, Vienna, and Wittenberg. *Henry the Fifth* contains many names, as is fitting for a story about an English king and his contemporary monarchs. Several names are part of aristocratic titles yet still generate a geographic identification. They include Agincourt, Bedford, Berri, Brabant, Brittany, Cambridge, Canterbury, Crete, Dover, Exeter, Iceland, Elbe, Ely, England, France, Germany, Lorraine, Masham, Meissen, Northumberland, Orleans, Paris, Poland, Saale, Scotland, Southampton, Wales, and Westmorland. Adjectival forms of country or other place names are common when referring to people from various areas.

MUSIC

Classical Music

Place names are in the titles of many classical musical pieces. The foreword to the book *Subject Guide to Classical Instrumental Music* by Jennifer Goodenberger says, "For centuries composers have drawn inspiration for their compositions in extra-musical sources and in the world around them."[6] In referring to the world, the author includes place names, and the book subsequently mentions a number of countries, cities, provinces, and features. Another publication, the 1975 edition of *Popular Titles and Subtitles of Musical Compositions* by Freda Pastor Berkowitz,[7] also mentions place names and music, principally classical.

Place names as titles bring to musical compositions a sense that the music has some relationship to a place and thereby possesses a special quality. Antonin Dvořák's piano duet, opus 68, was entitled "From the Bohemian Forest" and without question showed the composer's desire to translate his observations of the named feature into musical form. In his "Bells of Zlonice" (a town in Bohemia, a part of historical Czechoslovakia), Dvořák attempted to capture the nature of that place. Bedrich Smetana wrote "Ma Vlast" ("My Country"), a piece of six movements that reflected his feelings about places in Bohemia. One is a symphonic poem called "Vlatava" (in German, "Die Moldau"), which captures the nature of a major river with that name in the Czech Republic that rises in the mountains near České Budéjovice (German, "Budweis"), passes Plzeň (German, "Pilsen") and empties into the Elbe River along the border between the Czech Republic and Germany. The music realistically describes the river's characteristics from its beginnings as a small stream, then coursing over rapids, and finally passing slowly through flat areas. Adding a further note of nature, the last part also indicates a rising moon.

In "Bachianas Brasileiras" for voice, piano, and/or orchestra, Heitor Villa Lobos referred to five places in Brazil. The main theme of his Symphony no. 6, "Montanhas do Brasil," reflected the profile of a mountain near Rio de Janeiro that he traced on graph paper.[8] Carl Nielsen wrote an orchestral work entitled "Journey to the Faroe Islands." J. S. Bach wrote six pieces for the keyboard entitled "English Suites" and another six called the "French Suites." In his "Scenes Alsaciennes" for orchestra, Darius Milhaud named five movements after French provinces. Felix Mendelssohn traveled to Edinburgh and was so impressed with the area that he composed his third symphony there and called it the "Scotch Symphony." A later trip to Italy resulted in his fourth work, the "Italian Symphony." Wolfgang Amadeus Mozart visited Paris, where he composed Symphony no. 31, the "Paris Symphony." His Symphony no. 36, the "Linz Symphony," and Symphony no. 38, the "Prague Symphony," relate to those cities. Sergei Prokofiev wrote his "Stalingrad Sonata" during World War II, when that city was besieged by the German army. While never there, Ralph Vaughan Williams composed "Antarctica," his seventh symphony, with instrumentation, voices, and wind sounds to replicate what he felt must be the environment of that continent. It was written for the film *Scott of the Antarctic.*

Johann Sebastian Bach's "Brandenburg Concertos," while honoring his sponsor, the prince of Brandenburg, also indirectly relate to the administrative area of Germany with the same name. Johannes Brahms named his second symphony the "Wörthersee" to capture the landscape around a lake of that name in southern Austria where he resided for a period of time. He also wrote a number of Hungarian rhapsodies with melodies reminiscent of songs he heard in that country. Franz Joseph Haydn's Symphony no. 104 is the "London Symphony." He and Franz Liszt also composed pieces connected with Hungary. In Robert Schumann's Symphony no. 3, "Rhenish," he tried to capture the scenery of the Rhine area as a result of a journey he took to visit Cologne (located in Germany, where its name is Köln). Perhaps the most recent piece of music having a national tie is "An Israel Symphony" composed by Arnold Saltzman, a cantor of the Adas Israel Congregation in Washington, D.C. The premier performance was there in June 1998.

Aaron Copland's music about American landscapes has contributed greatly to an appreciation of the nature and beauty of areas in our country. His *Appalachian Spring* and *Grand Canyon Suite* in many ways are theme songs of the named areas.

Sometimes, however, references to places were not always the action of composers. Persons or publishers familiar with composers on some occasions gave a place name to a title for various reasons, one of them being to make the compositions commercially popular with audiences in cited places. Haydn's "Italian Overture" is an example.

Popular Music

Popular songs, particularly those composed in the United States, are listed according to places and a number of other themes in *The Green Book of Songs by Subject*, by Jeff Green.[9] The statement, noted above, that composers find inspiration in extramusical sources and in the world around them has perhaps even greater application to popular music. Titles and lyrics for this form of music have drawn on a wide variety of ideas and associations. In most cases, place names are used to evoke emotions of romance or a sense of affiliation with a specific place or a natural feature. Names known mainly to local residents will generally make the song popular among them, but there are instances where the appeal of the music will also serve to make the feature known to many others. Songs may also contain names because the composer or lyricist feels they impart an appealing geographic message or provide sounds that generate interesting or even comical audial characteristics. Without question, the use of names also is seen as a device for selling music.

Green's book identifies names of some 900 U.S. cities or places in the titles, lyrics, or both, of popular songs, and the names range from Abilene to Woodstock. All fifty states are also mentioned, among the most frequent being Alabama, Kentucky, Louisiana, Tennessee, and Texas. The book refers to Carolina or the Carolinas some thirty times, which probably means that such names include North Carolina (listed only five times) and South Carolina (listed only once). On the other hand, Massachusetts is mentioned only four times, and Utah five times. Western or country music is, clearly, associated more with certain states.

Places in the United States and elsewhere were common in popular songs written in the United States for many years. Since the 1960s, however, entertainment modes have radically changed, and other themes seem to have gained popularity. As to songs with names, people born before 1950 may easily recall such titles as:

"Arrivederci, Roma"
"By the Banks of the Wabash"
"California, Here I Come"
"Carry Me Back to Old Virginny"
"Dixie"
"I Left My Heart in San Francisco"
"Moon over Miami"
"Moonlight in Vermont"
"Carolina in the Morning"
"On the Road to Mandalay"
"The Beautiful Ohio (River)"
"The Red River Valley"

Many popular songs commemorate England, Scotland, and Ireland. Two famous pieces are "By Yon Bonnie Banks" (Scotland) and "When Irish Eyes Are Smiling." The song by Johann Strauss Jr. about the principal river in Austria, "The Blue Danube," is well known far beyond that country.

Music popular in Canada also relies on place names. For example, in the province of Quebec, there are some thirty-five songs that contain names of places.[10] Without question, other countries also have songs with names.

Songs also use names for lyrical or comical purposes. Two examples are "I Got a Gal in Kalamazoo" and "Chattanooga Choo-choo" A hit song of the 1950s was "Take Me Back to Constantinople," which gives its current name, Istanbul, and the reasons for the change.

Special Music

Perhaps the greatest number of names in any popular song occurs in "I've Been Everywhere," a piece composed in 1962 by Geoff Mack in which a hitchhiker mentions ninety-nine places he had visited (all but three being in the United States). Their locations do not represent any logical pattern of travel, but the names are chosen to permit a degree of rhyming. Samples illustrate the pattern. The first verse begins with the following names: Reno, Chicago, Fargo, Minnesota, Buffalo, Toronto, Winslow, Sarasota. The first names of another verse are Pittsburgh, Parkersburg, Gravellburg, Colorado, Ellensburg, Rexburg, Vicksburg, Eldorado. While included in the music to indicate the extent of the hitchhiker's travels, they nevertheless evoke associations with the named places and give the song an interesting flavor.

An unusual song with place names is the "Geographical Fugue," an a capella number written in 1930 by the German composer Ernst Toch. Its verse contains names of twelve places and features around the world. The performers (male and female voices) occasionally sing several names in unison and in varying sequences; at other times some singers pronounce a name slowly while others sing a different name more rapidly for a contrapuntal effect. The song has musical bars and time indicators, but there are no notes otherwise except to indicate fixed levels of tones at higher or lower pitches. After migrating to the United States from Germany, Toch changed some names to ones more easily pronounceable in English than in the original German. For example, the name Ratibor (a town then in Germany and now in Poland with the name Racibórz), which originally started the song, was changed to Trinidad. The reason was that its initial letter *r* of Ratibor in German has a trill the composer wanted to preserve. But a similar trill is more difficult in English, and Trinidad was deemed phonetically more suitable. The words of the song are "Trinidad and the big Mississippi and the town Honolulu and the lake Titicaca. The Popocatapetl is not in Canada, rather in Mexico. Canada, Malaga, Rimini, Brindisi, and Tibet, Nagasaki, Yokohama."[11] While the song gives no particular message, the musical presentation is both unique and interesting.

An example of songs only with place names is a series composed in 1985 by Theresa Sullivan Ranson, an elementary school music teacher in Richmond, Virginia. Called "Map Reading Song," it actually includes thirteen songs that feature the names of all nations of the world and their capitals, the states in the United States and their capitals, eleven regions (such as Central America and Micronesia), and seventy world water bodies. Mrs. Ranson wrote the lyrics because she believed all students should have some knowledge of geography and maps. Her fifth-grade students, numbering about 100, learn the names by heart and sing to her piano accompaniment. Known as the Globemasters, they have sung at various schools in this country and abroad. They performed at a program in Washington, D.C., celebrating the 200th meeting of the U.S. Board on Geographic Names in April 1995. The entire presentation takes about twenty-five minutes and mentions upward of 560 names.[12] With country names changing, periodic revisions have been essential. Unquestionably, the piece contains the most place names of any song.

Patriotic Songs

Place names in music are sometimes used in national anthems and patriotic songs. The book *National Anthems of the World*[13] provides anthems for 182 nations, including those that gained or regained independence after the end of the Cold War. The book has the music as well as English translations of verses as necessary. Curiously, nearly half of the anthems do not contain the name of the nation or the associated adjectival form. This situation can be attributed to several factors. Perhaps the main reason is that when an anthem and its verses were composed, the specific country may not have had a

name. Further, many songs call attention to national and cultural characteristics that were seen as sufficient for the patriotic purpose of an anthem. For
example, "The Star Spangled Banner" does not mention the United States of
America. The song "America the Beautiful" mentions America in every
verse. Other patriotic songs may not specify America, but there is little doubt
as to the subject of the message. Similarly, the French and British anthems
do not cite the names of those countries. On the other hand, the Canadian
anthem begins "O Canada." In times of war, songs refer to one's own country, the enemy's country, and places where soldiers may be stationed.
Among the songs written during World War I are the British melodies "It's
a Long Way to Tipperary" and "Mademoiselle from Armentïéres." During
World War II, a very popular English song was "The White Cliffs of Dover."

Religious Songs

Religious songs that refer to places are principally in the category of Christian hymns, a number of which contain such names as Israel, Jerusalem,
Zion, Bethlehem, and Nazareth. Christmas carols often mention these
places. The 1978 hymnal of the United Church of Christ includes numerous
references to these names and others. One hymn refers to the delta (of the
Nile River), the Nile, and Egypt. This book and those of other denomina-
tions also mention Africa, the Babylonian rivers, Bethel, Calvary, Canaan,
Ceylon, the city of David, Galilee, Gethsemane, Greenland, India, and the
Jordan. Most of these names served principally to remind singers that the
places are essential parts of Christianity, but some are referred to for the
sake of analogy. Use of such names helps communicate the important connections between events and places.

Songs in Areas without Written Languages

In the peninsular part of Malaysia are the Temiars whose traditions are
perhaps centuries old. Through songs they transmit information related to
their forebears, their life customs, the nature of their territories—including
the names of features—their health status and other personal and com-
munal circumstances.[14]

PAINTINGS AND PHOTOGRAPHS

Landscapes are common elements of paintings and photographs, and in
many cases the title of a picture will include the name either of the specific
feature portrayed or of the associated area. Among highly regarded landscape paintings are the works of Claude Monet, Camille Pissarro, Pierre August Renoir, Alfred Sisley, and other impressionist artists who portrayed the

Seine River and its surroundings in the late 1800s. The Phillips Collection, an art gallery in Washington, D.C., had an exhibition in late 1996 and early 1997 called "Impressionists on the Seine." Of the sixty paintings, fifty-four portrayed places associated with Paris and a dozen or more other towns on the Seine, mainly downstream from Paris. Some scenes covered small areas while others portrayed extensive landscapes associated with the Seine. Most titles give the name of the river as well as of a specific feature such as a settlement or a region. Other titles identify the river and an activity of people such as boating or enjoying a picnic. Titles include *On the Bank of the Seine, Bennecourt* by Monet, *The Seine at Port Marly* by Pissarro, *Boating on the Seine* by Renoir, and *Banks of the Seine at Argenteuil* by Sisley. The popularity of the Seine as a subject of art reflects its importance as a geographic feature that has served as a transportation artery, an economic resource, and a place of enjoyment and recreation. Similar artistic portrayals characterize popular natural features in virtually all countries.

In some cases, an artist may have been one of the few people to have seen the place portrayed, but with time the location became increasingly known by others. Whether a title mentions an area or a feature known to a few or to many, pictures most likely show landscapes that may appear very differently to viewers of today. Thus *Chadds Ford Landscape*, a beautiful painting by W. C. Wyeth, shows a rural scene in Pennsylvania with fields, trees, a farmhouse and outbuildings, and hills near a place with that identifying name. Most likely the scene has changed dramatically since it was painted in 1909. Winslow Homer painted numerous pictures of rivers, coastal areas, seascapes, and rural places, giving each its local name. Included are areas in New England, the Caribbean, and England. Without question, he captured the realities of the areas and associated them with the appropriate names. Jan Vermeer, while not known to have traveled extensively beyond his native Holland, painted scenes of Delft and surrounding areas. Interestingly, he also had an appreciation of cartography and geography, for many of his pictures of people in residential settings feature a map or a globe in the background.

Landscapes and natural features also are common subjects for amateur and professional photographers. Travelers often send picture postcards of landscapes with a message, "Having a good time. Wish you were here." Of professional photographers, Ansel Adams created an outstanding reputation for his impressive photographs of landscapes, which he identified with names of depicted features. Perhaps the best known is titled "Half Dome, Thunder Cloud," taken in Yosemite National Park. In the introduction to his book of Adams's pictures, John Szarkowski notes that Adams produced "technically flawless photographs of magnificent natural landscapes."[15] Another photographer, Edward Steichen, photographed landscapes, buildings, vegetation, and other outdoor scenes, each of which carried an identifying place name.

NOTES

1. Marcel Proust, *Remembrance of Things Past* (New York: Random House, 1934), 2:296.

2. Proust, *Remembrance of Things Past*, 2:720.

3. Kathleen Norris, *Dakota: A Spiritual Geography* (New York: Ticknor & Fields, 1993).

4. Garrison Keillor, *Lake Wobegon Days* (New York: Viking Penguin, 1985).

5. Stephen Vincent Benét, "American Names," in *Devil and Daniel Webster and Other Writings* (New York: Penguin), 1927. Renewed 1955 by Rosemary Carr Benét.

6. Jennifer Goodenberger, *Subject Guide to Classical Instrumental Music* (Metuchen, N.J.: Scarecrow, 1989).

7. Freda Pastor Berkowitz, *Popular Titles and Subtitles of Musical Compositions*, 2d ed. (Metuchen, N.J.: Scarecrow, 1975).

8. Vasco Mariz, *Heitor Villa-Lobos: Latin American Monographs* (Gainesville: University of Florida Press, 1963), 46.

9. Jeff Green, *The Green Book of Songs by Subject: The Thematic Guide to Popular Music* (Nashville: Professional Desk References, 1995).

10. Henri Dorion, former chairman of the Québec Commission de Toponymie, personal correspondence, 1995.

11. Ernst Toch, *Geographical Fugue* (Miami, Warner Bros. Publications, U.S. Inc.).

12. Theresa Sullivan Ranson, phone conversation with author, 20 October 1995.

13. W. L. Reed and M. J. Bristow, eds., *National Anthems of the World*, 8th ed. (London: Cassell, 1993). The book lists 182 "nations" but does not define their political status as fully independent and sovereign territories. For example, Wales is included. Thus the publication does not necessarily conform to U.S. or U.N. guidelines as to national terminology. It does, however, recognize that changing political conditions will continue to have an effect on national anthems.

14. Marina Roseman, "Singers of the Landscape," *American Anthropologist* 100, no. 2 (1998): 106–121.

15. Tim Hill, ed. *The Portfolios of Ansel Adams* (Boston: Little, Brown, 1981).

III

Place Names
Are Not Permanent

The next two chapters note how place names are often modified or entirely replaced by a variety of processes, including political decisions, ethnic groups moving from one area to another, military events, or a perceived need by populations to restore earlier names. The impact of writing systems on names and on communications of many kinds is documented.

6

Where in the World
Is That Place?

NAMES IN THE HEADLINES

Reporters always ask "where?" as well as "who, what, when, how?" News stories and even headlines include names to pinpoint nations or cities where described actions take place. Oftentimes, given the nature of the world today, the places cited are new or unfamiliar. When they appear in the headlines for the first time, people ask, "Where in the world is that place?"

THE IRON CURTAIN IS LIFTED

Despite stated solidarity between the Soviet Union and its allied countries following the end of World War II, the alliance had weaknesses that eventually led to its collapse. Perhaps the most critical event in the disintegration of the alliance was the fall of the Berlin Wall in 1989. The ensuing separation of East Germany (its legal name was Deutsche Demokratische Republik) from the communist bloc captured the attention of the world and served to initiate similar actions by other segments of the Soviet Union.

The Polish People's Republic soon became an independent state, changing its name to the Republic of Poland.[1] With the further decline of political power of the former Soviet Union, other countries once affiliated with that nation also changed their governments and their names. They retained traditional elements of their names but abandoned terms such as "socialist," "democratic," and "people's," which were recognized as identifiers of communist forms of government. The People's Socialist Republic of Albania became the Republic of Albania. The People's Republic of Bulgaria became the Republic of Bulgaria. The Czechoslovak Socialist Republic became the Czechoslovak Federative Republic. Within a short time, however, the Slovakian citizenry expressed concern about their status in the new structure, and the name was changed to the Czech and Slovak Federative Republic. In 1993, the units separated to become two nations, the Czech Republic and the Slovak Republic. The Slovak Republic recognized Slovakia as its short-form name, and several years later the Czech Republic adopted Czechia as its short-form name. The Socialist Republic of Romania became Romania.[2]

The Baltic countries of Estonia, Latvia, and Lithuania claimed headline attention for several reasons. Although these entities were absorbed into the Soviet Union and were proclaimed republics of that nation after World War II, the United States never officially recognized the Soviet occupation. With the political changes that rocked Eastern Europe in the years beginning in 1989, these nations reclaimed their independence as the Republic of Estonia, Republic of Latvia, and Republic of Lithuania, respectively.

With the collapse of the Soviet Union, the other Soviet republics declared their independence. The Russian Soviet Federative Socialist Republic became the Russian Federation. In 1991, the former republics of the Soviet Union formed the Commonwealth of Independent States to coordinate relations between the new entities.

Within Russia, other changes were taking place as well. In 1991, the citizens of Leningrad voted to take back the city's former name, St. Petersburg (or Petrograd, the name based on the official Russian Cyrillic spelling). Many people, nevertheless, wanted to retain Leningrad because of its fame as a Soviet bastion staving off German attacks during World War II. Interestingly, local people decided to keep Leningrad as the name of the district in which St. Petersburg is located. By 1993, all fifteen of the former republics of the Soviet Union had become independent countries. As part of this process, local authorities decided to change not only their country names but also names of cities, provinces, parks, roads, buildings, and other features that honored heroes of the communist era. The changes were manifest. In Ukraine alone, more than 90 percent of the names were scheduled to be altered. In many cases, former place names were brought back.

THE FORMER YUGOSLAVIA

Also related to the collapse of the Soviet Union were changes affecting what was once the Socialist Federal Republic of Yugoslavia. Several former republics of Yugoslavia became independent countries. While Serbia and Montenegro assert they are one republic known as Yugoslavia and while there is general recognition of that claim, the United States officially does not recognize such a state but continues to see those entities as two separate bodies.

Ethnic rivalries in the former Yugoslavia remain. During much of 1999, an administrative district of Serbia—Kosovo—was virtually the only foreign area to be mentioned in the press and TV media. A majority of the people in that district are ethnically Albanian, a situation dating from the days prior to the Ottoman Empire's jurisdiction over much of what has been termed the Balkans. For the first part of 1999, military forces representing Serbia, the principal republic of what is locally called Yugoslavia, occupied much of Kosovo in an attempt to defeat the "Kosovo Liberation Army." Stating it represented the Albanian population of the province, that army's goal was to create an independent nation. In mid-1999, after substantial attacks by air units of the North Atlantic Treaty Organization (NATO), the leaders of Serbia agreed to cease military action in Kosovo and to pursue a peaceful resolution of problems.[3]

Earlier in the 1990s, people in the neighboring state, Bosnia and Herzegovina (the official U.S. name for the area formerly called the Yugoslav Republic of Bosnia and Herzegovina), created territorial disputes. The problem was that its differing ethnic factions—the Croatians, Moslems, and Serbs—had claims to parts of the new country that could not be peaceably settled.

In 1995, the United States hosted a conference at Dayton, Ohio, where representatives of the concerned factions and other nations developed a general agreement as to the political future of Bosnia and Herzegovina. The National Imagery and Mapping Agency prepared a series of large-scale maps that showed topographic details, transportation routes, cities and towns, place names, and proposed boundaries that the parties could use to define the extent of the recommended sovereign areas. The representatives generally agreed that the map provided logical territories for future national occupation. Even the experienced observer of international boundaries will wonder, however, how the new territories could ever provide satisfactory homelands for the cited cultural groups. The overall entity is broken into numerous areas that are sometimes separated from each other as exclaves and have very complicated borders. While no settlement has been reached, a natural question arises: What names will be chosen by the different political units should they gain independence? Given the delicate cultural and political balance of this and other parts of the Balkans, it appears likely that similar disputes—and names issues—will arise elsewhere in the region.

OTHER COUNTRIES CHANGED NAMES

In addition, several countries not contiguous or even close to the former Soviet Union dropped the words "socialist" or "people's" from their formal titles. In Africa, the Socialist Republic of Benin became the Republic of Benin and the Socialist Republic of Mozambique became the Republic of Mozambique. At the same time, such countries took steps to eliminate governmental structures associated with the former names. Another merging of countries took place at the southwestern corner of the Arabian Peninsula after years of political and territorial confrontation. North and South Yemen, formally called the Yemen Arab Republic and the People's Democratic Republic of Yemen, respectively, merged as a single country, the Republic of Yemen.

NAMES IN THE MIDDLE EAST

The names Israel, Lebanon, Syria, and Palestine are often in the headlines because of conflicts regarding territorial rights—or at least concepts of such rights. The United States does not recognize Palestine as a nation, but such an entity is represented by the Palestine Liberation Organization (PLO). The precise limits of territory claimed by the PLO are disputed by Israel. Virtually all of it extends westward from the Jordan River and the Dead Sea as two bulges along the eastern boundary of Israel. The United States calls this the West Bank. For many years, Israel has called this area Judea and Samaria and has expressed the view that since it was part of ancient Israel, it should remain a part of Israel. Negotiations as to the political future of this territory remain under discussion. Another area is the Gaza Strip (a narrow piece of land adjacent to Egypt along the Mediterranean Sea), which was granted self-rule in 1994, although it was under dispute by Israel and the PLO for many years. Another name in the headlines is the Golan Heights, a high plateau in Syria that overlooks the northwestern part of Israel and is occupied by Israel. In July 1999, Israel participated in discussions with the United States concerning the return of the Golan Heights to Syria but no conclusions were reached. Discussions between representatives of Israel and Palestine hosted by President Clinton at Camp David in the United States in August 2000 concerning sovereignty issues also produced no conclusions.

NAMES IN SOUTHEAST ASIA

Although modifications of national names can reflect political circumstances, names can change because of cultural factors. The government in Burma took a new name, Myanma, in 1989 to reflect a term defining an ethnic group that constituted a majority of the population, despite the fact that the name Burma related to the name of another important ethnic group. The official announcement received by the United States stated that as of a specified date, Myanma would be the "English-language" name. This statement was puzzling because what actually was meant was that it was the Roman-alphabet version of the name as written in the Burmese script. A further point of confusion materialized shortly thereafter when another announcement said the letter *r* would be added to the end of the name for better pronunciation. This raised the question of how that letter would be pronounced. In any case, the local official spelling is now Myanmar. The official position of the United States, however, is to retain the name Burma.

Cambodia, a country neighboring Burma (or Myanmar), also changed its name. That state several years ago proclaimed that its name in the Roman alphabet would be Kampuchea. This change is also related to a resident ethnic group, but the spelling is actually a variation of the name Cambodia. Local authorities decided Kampuchea when pronounced was phonetically more acceptable. While recognizing the name Kampuchea for formal purposes, the United States uses the name Cambodia for most communications.

My Lai, Vietnam—Was It My Lai 1, My Lai 2, My Lai 3, My Lai 4, My Lai 5, or My Lai 6?

For the past three decades, the "My Lai Massacre" has received much press attention. The killing of innocent civilians by American troops in 1968 in an area in Vietnam generated investigations of local military operations and subsequent trials. A platoon leader, Lt. William Calley, was convicted as the principal military officer responsible for the deeds. Of several books on the subject, a recent one is *Facing My Lai: Moving Beyond the Massacre.*[4] As the issue became more highly publicized in the early part of the 1970s, the staff of the Foreign Names Committee of the U.S. Board on Geographic Names at NIMA raised a question of where exactly the crime occurred. Their knowledge of place names in Vietnam and in other nearby countries indicated that several settlements could have basically the same name. This pattern reflected a cultural habit. Young people moving from their birth places and setting up a community elsewhere would transfer the same name to retain tribal identification, but they would add a number. The BGN staff said there were as many as twelve places called My Lai. Questions followed. Although a hamlet called My Lai was the scene of horrendous crimes in 1968, which of several places was involved? Was another My Lai indeed the target, since it had heavily armed Vietnamese troops and civilians who were combat ready? Did local American commanders misinterpret maps, which did show six hamlets called My Lai?[5] There was no official request for the staff to pursue this question, so no further research ensued and references to a single My Lai continued. While the crimes cannot be discounted to any degree, the existence of virtually identical place names in areas of military conflict must require careful analysis to avoid possible misconduct of operations. In this case, according to official U.S. records, it is unlikely that American troops attacked the wrong place.

THE PERSIAN GULF OR THE ARABIAN GULF

During the Gulf War in 1991–1992, the name "Persian Gulf" was frequently in the headlines. A question arose concerning that name. Although it had been in use for many years because of its association with Persia,[6] U.S. military commanders in the area requested the name be changed to Arabian Gulf. The recommendation had some logic. The name Persia was no longer a country name, having been replaced in 1920 by the name Iran.[7] U.S. relations with Iran were not good. And U.S. allies along the western bank of the gulf generally were opposed to Iran and did not use the name Persian Gulf. In fact, some of those countries did not accept incoming U.S. mail if any addresses referred to the Persian Gulf.

The question came before the BGN Foreign Names Committee. Members representing the Central Intelligence Agency, the Department of Defense, the Department of State, and the Library of Congress carefully studied the matter. A major BGN principle is that no country can change a traditional name of any feature whose territory is beyond the sovereignty of a single nation. The request, however attractive under the circumstances, was turned down. Another argument against the change was the existence of a nearby body of water already named the Arabian Sea. It extends between the eastern shore of India and the southern coasts of nations on the Arabian Peninsula to the west. The Persian Gulf connects with the Arabian Sea by way of the smaller Gulf of Oman, and confusion would have been generated by having two major bodies of water with the name "Arabian."

A CASE OF NAMES IN HEADLINES OF THE PAST

Groups in Austria continue to express concern about place names in the Italian province of Alto Adige. Prior to World War I, this area was part of the Austro-Hungarian Empire and was called Süd Tirol. The Allies gave the territory to Italy as a reward for that country joining them in the war against Germany and the Austro-Hungarian Empire. The fact that a mountain chain effectively separated the province from the empire was cited as evidence that it should be part of Italy. Allowing Austria (the relevant country existing after the empire was dissolved) to keep Süd Tirol would give that country an unfair strategic advantage over Italy in case of a future war. Furthermore, Italian authorities claimed that the region at one time was inhabited by Italian people. In any case, when Italy took possession of the area, a program was started to change place names to Italian versions. The capital, Bozen, for example, became Bolzano. Changes also affected names of streets, hotels, and even people's names on tombstones. The reaction by local Austrians was predictably one of anger and frustration. For several years, a group of Austrians has attempted to restore what they consider time-honored and legitimate names in an area they claim as ancestral territory. Representatives of Italy and Austria have studied the question but as of late 1999, there was no evidence of a resolution in favor of the Austrian position.

A discussion of new and changing country names benefits from the inclusion of appendix B, a list of current country names of the world and related information produced by the Department of State in January 2000. It gives the short-form and the long-form name of each country (where appropriate). Both forms are spelled in accordance with BGN decisions and, where applicable, reflect its rules to convert spellings from non-Roman-alphabet writing systems to forms in the Roman alphabet. Each country also has a two-letter code assigned by the U.S. National Institute of Standards and Technology for general use throughout the U.S. government. Using codes eliminates the need for full national names in various communications when their spellings can be cumbersome. In addition, the list shows the capital of each country and, where appropriate, romanized versions of names.

NAMES IN FUTURE HEADLINES

People living in northern Italy in the Po River basin exercised the principle of self-determination in 1996 by proclaiming a territory with the name Padania as an independent sovereign republic. This act was evidently a manifestation of long-realized differences among people living in the northern and the southern parts of Italy. Apparently no further efforts have been made since the initial proclamation.

The complex cultural nature of India, with sixteen official languages, has given rise to nearly as many groups seeking greater independence. With time, political actions may see the country separate into a number of independent nations. In Latin America, indigenous populations in the Andes have striven for recognition as an independent territory coinciding with the area where they live. A name for this entity would certainly be news.

The next chapter, "Names in Multilingual Countries," describes situations is some countries that could also lead to names in future headlines.

NOTES

1. This chapter refers principally to conventional short-form names, as opposed to conventional long-form names or local official names. Poland is a conventional short-form name, while the conventional long-form name is the Republic of Poland. Chapter 11, "The Terminology of Names," gives more details on this aspect of names.

2. Curiously, Romania is currently the only name. There is no other formal or conventional short-form name.

3. In May 1999, NATO aircraft bombed a building in Belgrade, the capital of Serbia, believing the structure had a military function, as indicated by aeronautical charts. It was actually the embassy of China, and the bombing error incident generated international comment. Prior to the attack it had become the Chinese embassy but unfortunately cartographers did not have correct documentation to update charts. Although the case did not involve a place name, the fact is that having a wrong building name is similar to having an incorrect place name to identify a target designation. Chapter 17, "The North Atlantic Treaty Organization," gives additional information about NATO.

4. David L. Anderson, ed., *Facing My Lai: Moving Beyond the Massacre* (Lawrence: University Press of Kansas, 1998). The publication contains commentary of some forty individuals, including former soldiers on the scene and military officers later involved in court martial proceedings.

5. The Center for Military History of the U.S. National Archives at College Park, Maryland, has extensive files on My Lai. Reports with maps of military actions refer to My Lai 1, My Lai 2, My Lai 3, My Lai 4, My Lai 5, and My Lai 6. While most publications refer only to the name "My Lai," there is evidence that My Lai 4 was the center of action.

6. Features that are not completely within the jurisdiction of one country or that fall within two or more countries are sometimes called "international features." Countries having sovereignty over parts of international features can name such parts. In the case of the Persian Gulf, the influence of Persia in the area over the centuries gave birth to that name, at least in the English language. The long existence of the name has resulted in its widespread acceptance.

7. Iran took no action to modify the preexisting name.

7

Names in Multilingual Countries

DIFFERENT LANGUAGES, DIFFERENT NAMES

Countries with populations speaking more than one language most likely will have different names for a number of places.[1] The names may be basically different or their spellings may vary according to the pertinent language. In countries having languages other than the one or more officially recognized, people may also seek approval of their individual languages and of any subsequently different place names that are part of their culture. The existence of languages without corresponding writing systems further complicates efforts to recognize local names. These factors can frustrate what has often been called the main goal of names standardization: each named feature in a country should have only a single name. Features with two or more names confuse people within a country and elsewhere seeking the locations of places. This situation is relatively widespread, since there are few countries with only one language. Another complication of spelling names is related to writing systems applied to them. This topic is described in chapter 12, "If Your Language Is English, How Do You Write or Pronounce Arabic or Russian Names?"

HOW MANY LANGUAGES ARE THERE?

It is difficult, if not impossible, to calculate the precise number of languages because some are associated with only a small number of people or may be a dialect of another, or are only spoken. One source has numerous maps and lists of names indicating there are nearly 8,000 languages and subgroups.[2] The continent of Africa has upward of 1,000. New Guinea has about 700. Native Americans in the United States speak some fifty different languages. Even more languages existed in the past, but the growth of international communications and other social and ethnic changes have caused many to become extinct.

The relationship between languages and place names is demonstrably clear. In regard to giving names to places, a major factor is whether the local population can write. If not, there is no written record of names. A related point is whether such populations have given place names (as they are generally understood to be) to specific features. It is possible that in such areas the people refer to a feature by a generic term (i.e., river or lake). Even where populations are able to write, albeit to a limited extent, they still may refer to a feature only by a generic term.[3] Further, such names may be in local use but unknown to people living in other areas.

Many nations created or expanded their sovereign areas through military or political actions, and in the process often incorporated territories where people spoke different languages. Depending on various factors—including the dynamics of local languages and associated cultural situations—languages brought in by conquering states might be modified to conform with local practices. Conversely, languages of invaders might influence or even replace local ones. Furthermore, where people became isolated, their languages were subject to change and became radically different. These conditions have resulted in a complex pattern of languages, a situation that seems to be growing as efforts continue to recognize and preserve them.

EXAMPLES OF DIFFERING PLACE NAMES

Switzerland has four official languages, with 65 percent of the population speaking German, 18 percent speaking French, 12 percent speaking Italian, and 1 percent speaking Romansch. Reflecting this pattern, the country itself has four official names: Schweiz (German), Suisse (French), Svizzera (Italian), and Suiza (Romansch). Since Romansch is used by such a small percentage of the population, the capital of the country has no spelling in that language. Its three standard names are Bern (German), Berne (French), and Berna (Italian). Other features may have two or more commonly known names, but in the areas where one of the three languages predominates, the names generally conform to the local language. As one language area gives way to another, road signs usually display place names in two languages (or even three, if appropriate). The different languages in Switzerland are based on those used in the various political entities of the country, first beginning with the merging of three elements as a confederation in 1291 and continuing with the addition of other areas over the centuries.

In many parts of the world where areas were overtaken by foreign invaders, today's descendants of original populations often seek restoration of their original languages and place names. In Ireland there is a movement to restore Gaelic as the local language and to recognize relevant place names. A similar action in Wales brought official recognition of Welsh place names.

As of July 1990, the Soviet Union perhaps had more than 200 languages and dialects, of which at least 18 were spoken by 1 million or more people. Each of the fifteen republics had a separate native language, even though some were related. Most of the major languages also used a different writing system. Although the republics continued to use their own languages for many purposes, during the life of the Soviet Union, Russian was the official language and was uniformly used by those aspiring to careers at levels outside of their own republics. Similarly, the Russian version of the Cyrillic alphabet was the official writing system in the Soviet Union. With the collapse of that empire and with the official establishment of native languages and writing systems for the newly independent nations, countless place names were changed to reflect their historic nomenclatural roots.

The newly independent countries are intensely occupied with creating their national institutions and reestablishing their economies, and a major task has been to establish agencies responsible for names. Some countries are collaborating with the United States to obtain ideas concerning principles, policies, and procedures of name standardization established by the U.S. Board on Geographic Names.

Canada has two official languages, French and English. While French is largely restricted to the province of Quebec, all official Canadian documents must be printed in both languages. A number of people, especially in some eastern areas, are bilingual, and employees of the Canadian government are often required to be able to use both languages regardless of their place of employment. In addition, many other languages are spoken by a number of indigenous people (Indians and Eskimos) living in areas they inhabited prior to the advent of English and French and in some areas Russian settlers. The effect on place names is considerable. The province of Quebec has a names standardization body, Gouvernement du Québec Commission de Toponymie, with a program to assure names in that area reflect the French culture. Yet some names in Quebec remain English or have local Indian forms. Some major features such as rivers or lakes have valid names in English and French. Other Canadian provinces and territories have names authorities as well. While being responsible for decisions to approve names within their jurisdictions, they collaborate with the Canadian Geographical Names Board regarding a range of policies and procedures.

Various Canadian organizations are working with ethnic minorities to develop standard writing systems. Christian missionaries nearly 200 years ago developed syllabic systems that still have validity but have experienced changes. The task of stabilizing and expanding the systems is difficult, inasmuch as the many user groups have different heritages that tend to promote variations. The systems are based on a wide variety of non-letter graphics along with certain Roman letters. The Inuit people of Canada, which, although small in number, occupy extensive areas across the northeastern part of the country and have numerous languages. In 1999, Canada created a new territory called Nunavut in the northeastern part of the country whose area amounts to approximately 25 percent of the national total. The official language of residents is Uikitut. Writing systems developed by linguists for this and other minority languages have produced complicated spellings that include various diacritics and even non-letter elements. In the Yukon Territory, for example, such names are *Kweteni?aá, Xây Gûn, Nàgas' èX'àyi*, and *Nii'ii*. These kinds of names may satisfy language needs of local populations and be linguistically suitable, but it is unlikely they convey information to people elsewhere. Even with growing local usage, such writing systems, when applied to place names printed on maps or other geographical documents intended for wide distribution, may not be understood by a very high percentage of the Canadian population.

South Africa is looking at the names of many cities and other features that largely reflect the country's Dutch and English history. With the end of apartheid and the greater voice of different ethnic groups in national policies, will there be a new name for the country? Nineteen different languages are spoken in South Africa, of which eleven have official status. Will such names as Johannesburg and Pretoria be dropped in favor of new names or will each place have as many legal names as the local populations may desire? Already, Cape Town has five additional names: Ekapa, Kaapstad, Kapkaupunki, Kapstadt, and Le Cap. Authorities in that country are working to establish a names agency to deal with this challenging subject.[4]

India has sixteen different official languages, of which Hindi and English are predominant. In addition, twenty-four other languages are each used by a million or more people, and there are numerous other tongues, most of which are mutually unintelligible. The national illiteracy rate of 48 percent suggests that a number of the cited languages have no written form. The task of developing a list of standardized names is truly daunting. Recent name modifications receiving press attention include changing Bombay to Mumbai, and Madras to Chennai.

In the United States, the pattern is also dynamic. Early explorers and settlers spoke Spanish, French, English, German, and Dutch and, for many years, areas had populations speaking those languages. As noted before, the languages of these early groups influenced place names. While English is virtually the official language of the United States and is recognized as such by most states in the country, the languages of recent immigrants are becoming more common in areas where they are concentrated.[5]

For some time, the recognition of indigenous people brought pressure to reestablish what they recognize as their place names. In this regard, the names of two mountains periodically have received press attention over the years. Mount Rainier in the state of Washington is also called Tacoma by local populations and others who prefer what they claim is the original name. In Alaska, Mount McKinley has been called Denali, which is cited as the local Indian name. Otherwise, Indians and their supporters have moved to eliminate any names they find opprobrious that have been applied by others allegedly not familiar with their culture. Chapter 19, "Names in Dispute," discusses these and other cases.

As Hispanic populations in the United States increase, there are growing movements to spell place names of such origin to conform with Spanish customs. In other words, diacritics associated with Spanish place names should be retained. Similarly, the state of Hawaii asked that its native names be honored, complete with diacritics or special markings, and the U.S. Board on Geographic Names approved such action, at least as acceptable local variants. The Domestic Names Committee of the Board considers such cases at virtually every monthly meeting. At its July 1998 meeting, the committee reviewed and later accepted a recommendation of the Legislature of Guam to change the name of Agana, the capital of Guam, to Hagatña.[6]

Otherwise, the absence of writing systems in numerous areas raises the question of how to obtain the names of features if, in fact, there are any names. Even in areas having societies with written languages, recording feature names by practical procedures of fieldwork can be difficult.[7]

As communications advance at the local, regional, national, and international levels, detailed knowledge of languages (and writing systems) in many areas is required to record place names. Movements to recognize minority or even "dead" languages, and to create writing systems where none exist, necessarily involve place names. The increasing knowledge of a growing world of languages and the growing need to enhance a cultural awareness of local populations present sizable challenges to names programs.

NOTES

1. Note 1, chapter 1, "What Is a Place Name?" gives a definition of the word "language."

2. Christopher Moseley and R. E. Asher, eds., *Atlas of the World's Languages* (London: Routledge, 1994).

3. Chapter 11, "The Terminology of Names," discusses generic terms as well as other terms associated with place names.

4. The integration of ethnic groups in South Africa is having an impact on the management of names programs in the nation. A white South African who played an effective role in behalf of his country and at United Nations meetings for many years was replaced by a black representative at the session of the United Nations Group of Experts in January 2000. A paper distributed at the meeting indicated the change was made because the former representative did not adequately reflect the interests of the majority of that country's population.

5. The growing influx of Spanish-speaking immigrants has led to demands to give that language a status equal to English on a broad scale. Such recognition already exists in certain areas. Similar requests are expressed on behalf of other languages.

6. As is the case with most names having diacritics, pronunciation poses problems. Agana and Hagatňa have similar pronunciations, but people not familiar with that territory would probably be unable to say the new version.

7. Since 1987, the United States and the Pan American Institute of Geography and History have sponsored two-week courses each year (except 1988) in South and Central America on names standardization. Fieldwork in these courses requires various steps to locate names or verify their validity. These actions include interviewing local inhabitants, referring to existing maps and road signs, consulting with local officials, taking detailed notes, working with appropriate authorities to resolve differences in names data that can arise, and making full accounts of results for official records. In cases of unsettled problems, additional fieldwork may be required. A few countries elsewhere in the world are also developing similar courses.

IV

Efforts to End the Confusion

The next six chapters describe national and international programs developed to ensure the selection and application of correct names. One chapter gives attention to the U.S. Board on Geographic Names, the reason for its founding in 1890, the nature of its operations during times of national and global changes, and the manner in which its functions have contributed to standardization programs on a worldwide spectrum. In addition, there is information about other national programs and those of the United Nations as well. One significant topic concerns efforts to name undersea, Antarctic, and planetary features where there is no population. Another element concerns the creation of a special vocabulary related to place names. Attention also is given not only to the manner in which differing languages and writing systems have complicated efforts to resolve the confusion that characterizes so many names around the globe, but also to programs to treat relevant issues. A section also describes how publications called gazetteers have become essential records of place names. Appendices provide details to augment the contents of some chapters.

8

Countries Recognize
the Problem

THE SITUATION IN THE UNITED STATES

When it became an independent country in 1789, the United States began to explore territories west of the thirteen member states. In the process, the boundaries and areas of some of these states were redefined and new entities arose. Westward patterns of settlement, which included vast territories obtained by purchase or military campaigns, resulted in maps and records of previously unknown areas. The expeditions of Lewis and Clark, for example, generated many maps.

The names on maps and in reports came from many sources. Surveyors heard names as pronounced by local indigenous people, and they encountered other explorers or traders (many being French) who had given names to features. They then wrote the names as they thought correct. In addition, members of surveying parties sometimes named places. One complication was that several groups crossing specific areas one or more years apart might record or give different names for identical features. One group crossing northwestern regions, for example, might have seen a river in the spring when melting snow upstream caused soil erosion and gave the water reddish or brownish colors. Another group exploring areas to the south several years afterwards late in summer might have seen what they thought was a different stream, but it was really a southern extension of the same river when waters were low and appeared black. The first group might have called the feature the "Red River." The second, not knowing that earlier records had identified the feature, might have named it the "Black River." In addition, the U.S. Post Office frequently named local offices for individuals rather than for towns or settlements they served. This kind of inconsistent naming was widespread and struck many as illogical and confusing.

When federal officials compiled final reports and noted different names applied to segments of a single feature, they could not always resolve name differences. It was virtually impossible to send an expedition back to verify names and consequently officials selected a single name. The results often were less than satisfactory.

Another serious dilemma affected maritime operations. Navigational charts too commonly had wrong names for islands, bays, mouths of rivers, and ports in the United States and in foreign waters. The lack of correct names too often brought ships to dangerous places. Other countries were also affected by similar problems of inconsistent names on maps and charts.[1]

CREATION OF THE U.S. BOARD ON GEOGRAPHIC NAMES

By the latter part of the nineteenth century, these difficulties in the United States had grown to such proportions that a group of leading cartographers, geographers, academicians, and people in the U.S. government met to study the issue. They submitted a report early in 1890 calling for an official U.S. body to deal with place names.

After President Harrison studied the report, he issued the following executive order on 4 September 1890:[2]

> As it is desirable that uniform usage in regard to geographic nomenclature and orthography obtain throughout the Executive Department of the Government, and particularly upon maps and charts issued by the various Departments and Bureaus, I hereby constitute a Board on Geographic Names, and designate the following persons, who have heretofore cooperated for a similar purpose under the authority of the several Departments, Bureaus, and Institutions with which they are connected, as members of said Board.

The order then named individuals of the Coast and Geodetic Survey, Department of State, Treasury Department, War Department, Navy Department, Post Office Department, Smithsonian Institution, and the Geological Survey. These organizations were involved in a wide range of activities including exploration, surveying, map and chart production, navigation, maintenance of lighthouses, mail delivery, and diplomatic and military or related functions. The order continued:

To this Board shall be referred all unsettled questions concerning geographic names which arise in the Departments, and the decisions of the Board are to be accepted by these Departments as the standard authority in such matters.

The board's first report in 1892 defined its operational procedures and policies and reported on work carried out.

The board's mission changed with time. Until 1906, its functions were to examine duplicate or incorrect names appearing on maps, charts, or in other publications and decide on a single official name. In that year, the board gained additional authorization to consider any names recommended by employees of the government. Thus new names had to be submitted for review and approval before they could appear on any official map or chart. Furthermore, the name of the organization was changed to the Geographic Board but was altered again a few years later to the Board on Geographic Names. The board continued to work on its mission, with increasing attention to foreign names. Its published reports appeared at irregular intervals until the sixth report was published in 1932. It included events from the beginning until 1920 as well as several supplements published separately to cover decisions rendered for the years, ending in June 1932.

THE UNITED STATES ADDRESSES WRITING
SYSTEMS OF FOREIGN NAMES

One supplement in the sixth report was entitled "Rules for the Spelling of Foreign Geographic Names for Official Use in the United States of America." While the board had dealt with foreign names for some time, this was the first policy statement. For areas using the Roman alphabet, usage of locally spelled names was recommended unless there was an existing conventional English version. For languages with non-Roman letters, the guide called for application of the system of the British Royal Geographical Society. (Compared with later U.S. policies on foreign names, which cover twenty-seven language areas, the rules cited in the sixth report represented a useful but limited policy.)[3]

For a variety of reasons, the board's traditional programs declined between 1932 and 1943. It continued to work on names, but only a handful of people participated. The staff was reduced to two or three and reports of actions were infrequent.

MILITARY NEEDS REQUIRED THE REESTABLISHMENT
OF THE BOARD

The country's entrance into World War II in 1941 quickly demonstrated an urgent need for detailed maps of Japan, China, and other countries in the

Far East. As production of large-scale military maps of these areas began, it was soon evident that place names were lacking. Officials looking to the board for such information found that its small staff could not meet the challenges, and a plan was adopted to enlarge the body. Dr. Meredith F. Burrill, a geographer and former university professor serving with the Office of Land Management in the Department of the Interior, was appointed in 1943 as director of the department's newly formed Division of Geography. At the same time, he became executive secretary of the revived U.S. Board on Geographic Names. The working force soon grew to include about 176 geographers, linguists, cartographers, librarians, and others. While the program resembled earlier board efforts to provide names for all national agencies, the staff worked closely with army and navy personnel to assure the availability of names for maps and charts being produced to meet rapidly expanding military requirements. The body's work on domestic U.S. names remained at a relatively low level.

While associated with the board, the division's major task was to convert names on existing Chinese and Japanese maps from the scripts used by those countries to names spelled with Roman alphabet letters according to approved systems. During the war, romanized place names were also required for maps of other places. At war's end, upward of 7 million names had been processed, with a sizable majority for areas in the Far East.[4]

THE BOARD IS REORGANIZED

With the end of World War II, Congress began to reduce or eliminate programs related to military efforts. Many congressmen thought that the board was no longer needed, but others believed that peace was not firm and that military needs for maps and charts—and place names—would persist. The issue became controversial, and a number of hearings were held. As virtually the last action of the Congress in 1947, a one-vote majority resulted in the continuance of the board.

In the process, the structure and operations of the board were revised extensively to make it more responsive to the needs of military and civil agencies concerned with cartographic products. The board and its executive secretary remained attached to the Department of the Interior and reported to the secretary of that department. Members of the new board represented the Departments of Agriculture, Interior, State, army, air force, navy, and the Post Office and the Library of Congress. Representatives of these agencies met to spell out principles, policies, and procedures the board would follow. In the following years, the Department of Commerce, the Department of Defense, the Central Intelligence Agency, and the Government Printing Office were added. The interests of the army, air force, and navy were then represented by the Department of Defense. Basic decisions board members reached as they created operational guidelines included the creation of a Domestic Names Committee (responsible for making decisions on names in the United States and its territories and possessions) and a Foreign Names Committee (responsible for decisions about names in all other countries). The bylaws spelled out other provisions such as the election of committee chairmen, the creation of standing or temporary committees, and procedural matters as schedules for meetings, consideration of questions coming before the board and its committees, and distributing papers describing resulting actions.

As an interagency body, the board has an unusual structure. It does not have a separate budget. Members from cited agencies serve without compensation. The chairman of the board is a member appointed by the secretary of the interior for two-year terms. The executive secretary of the board traditionally also was the executive secretary for foreign names after that position was established in 1947 when the board was reorganized.

With time, the board experienced further changes. Action to standardize names did not always address the practical requirements to provide names to meet well-established map and chart production schedules. The National Mapping Division of the U.S. Geological Survey in the Department of the Interior needed names for its U.S. mapping programs and often felt the board, through its Domestic Names Committee, was not able to meet its production schedule. In 1958, the secretary of the interior transferred the staff of the board's Domestic Names Committee from the Division of Geography to the National Mapping Division to help assure that action on names responded to production schedules. The executive secretary for domestic names, a high-ranking person in the Department of the Interior, also pursued his functions in a manner that made the committee operate relatively independent of the board. Such a situation reflected the fact that the board was not subject to strict operational guidelines.

The change of 1958 produced an awkward situation: the staff of the Foreign Names Committee, whose program was considerably more extensive than that of the Domestic Names Committee, still remained within the Department of the Interior which otherwise had little if any responsibilities for mapping foreign areas. To resolve this dilemma, the staff moved in 1969 to the Army Map Service, a part of the Department of Defense. This agency was responsible for producing maps and relevant names for army purposes and earlier had its own names staff. Coordination with BGN procedures was required, but the liaison was not always assured. At the same time, the board's executive secretary was relocated in an element of the Defense Intelligence Agency in the Pentagon that coordinated all U.S. military cartographic programs, including that of the Army Map Service. The air force and navy also had units to work on names to meet their requirements.

In 1972, the Defense Mapping Agency was created to manage all military cartographic work, and the office of the executive secretary was moved to its headquarters. The Army Map Service was renamed the Defense Mapping Agency Hydrographic-Topographic Center, and its names staff expanded to include people from the former U.S. Navy Hydrographic Office who worked only on naval charts. The staff members were also merged with the staff of the board's Foreign Names Committee. The Defense Mapping Agency Aerospace Center, in St. Louis, retained a names staff, but increasingly the group at the Hydrographic-Topographic Center produced names to meet air force needs.

In 1996, elements of Central Intelligence Agency, the Defense Intelligence Agency, the National Aeronautical and Space Agency, the National Security Agency, and other organizations producing maps and related items from satellite imagery joined the Defense Mapping Agency to become the National Imagery and Mapping Agency. The staff supporting foreign names work of the Board on Geographic Names remained as part of the new agency.

In 1993, the position of executive secretary of the board was transferred to a person in the U.S. Geological Survey in the Department of the Interior who formerly had been serving as the executive secretary of the Domestic Names Committee and continues to do so. The executive secretary of the Foreign Names Committee and the staff supporting the board's foreign names work are at the National Imagery and Mapping Agency.[5]

Appendix C provides information about structure, membership, officers, and functions of the board and its committees. It also provides examples of committee actions on names proposals.

To meet a growing public demand for data about geographic names, the board and its support agencies have appropriate distribution policies. Appendix I identifies methods for obtaining information and publications about foreign and domestic names processed by the board.

OTHER NATIONAL AND INTERNATIONAL BODIES

The board has collaborated with numerous names organizations in the world, particularly those of Canada and, as noted, the United Kingdom. The Canadian Permanent Committee on Geographical Names (CPCGN), renamed the Geographical Names Board of Canada (GNBC) in 2000, was established in 1897 to deal with the host of new and difficult names in its vast provinces and territories, which were being explored and settled. The body prepared a set of procedures for general application, but recognized that the provinces and territories retained the right to approve names for their internal purposes. The procedures are rational and productive and the various publications issued by the GNBC and the provinces are impressive.

Over the years, a major challenge in Canada (and in other countries as well) has been the view that names used by native peoples should be recognized and published. As noted in chapter 7, some written versions of native names have forms that other people may find difficult to use. The status of French as an official language also has complicated national customs. While French names are limited mainly to the Province of Quebec, a number of names elsewhere officially have an English and a French form.

The Canadian authority also has a committee on undersea feature names, and in addition it forms groups to work with a variety of toponymic issues.

It is not involved with foreign areas to any particular degree, but consults with various Canadian bodies and the United States when information about foreign names is needed. Canada and the United States have worked together on a number of issues. They form the United States/Canada Division of the U.N. Group of Experts on Geographical Names and present common views when working with other experts associated with U.N. programs. Canadian participants in U.N. activities over the years have demonstrated admirable abilities to analyze and resolve challenging issues relating not only to Canada but also to U.N. programs. In addition, the United States and Canada have a joint committee whose goal is to eliminate any differing names of rivers or other features common to both countries.

The United Kingdom formed the Permanent Committee on Geographical Names for British Official Use (PCGN) in 1907. Since that time, U.K. experts have accomplished a remarkable record of compiling and publishing records of names (gazetteers) covering a number of other countries. The first gazetteers focused on members of the British Commonwealth, but eventually included other nations as well. These products generally set the pattern for BGN gazetteers published after 1947. Members and staff of PCGN also have played a strong role in many aspects of names standardization that has benefited toponymists of both nations for many years.[6] Although the United Kingdom is active in programs on foreign names—and is especially effective in U.N. names programs whose major mission is to encourage nations to create names agencies—the PCGN has no authority for names in the United Kingdom. This is the mission of the U.K. Ordnance Survey, which is responsible for national (versus foreign) mapping. The rationale for not having an agency to deal with domestic toponymy is that names in the United Kingdom have been stable for many years, and any responsibility for dealing with new names is that of the Ordnance Survey. PCGN has members from several offices involved with foreign affairs, international mapping, and private publishing. Its staff is housed in the Royal Geographical Society in London and works on a continuing basis according to directives established by the committee at annual meetings. Its schedule includes meeting with U.S. Board on Geographic Names every two years to discuss issues important to both sides. The meetings have produced bilateral agreements on names as they relate to the programs of the various international organizations. As noted in chapter 12, "If Your Language Is English, How Do You Write and Pronounce Arabic or Russian Names?" and chapter 13, "Gazetteers," the U.S./U.K. collaboration has been an unmatched model of international cooperation.

Many other countries, and the number is increasing, also have names agencies which, for the most part—and for good reason—are focused on domestic issues. Nevertheless, there is a growing amount of international collaboration that, for the most part, reflects U.N. recommendations. National standardization procedures, however, vary considerably. Some countries have a central body that lacks the authority to require all national agencies to collaborate. Too many nations have agencies, such as those responsible for transportation and census matters, that do not consult with each other regarding place names.

Curiously, some countries playing active roles in the United Nations or other international organizations concerned with place names have no central names authorities with clear-cut responsibilities. France, Italy, and the Netherlands are included in this category. Nevertheless, while having various kinds of problems concerning names, each has contributed to the development of useful principles. Other countries with active mapping programs also may have no names authorities. Several nations in Latin America have no offices, nor does Japan. Countries once part of the Soviet Union are now actively working to create national standardization authorities. A relict of the Cold War remains, however, in that North and South Korea have not been able to develop a common romanization system for their writing. It is indeed unfortunate that although many nations identify a national names authority, their work is so limited as to be virtually ineffective.

The U.S. Board on Geographic Names, through collaboration with the National Imagery and Mapping Agency and the Pan American Institute of Geography and History, inaugurated a program in 1987 to teach courses to Latin American countries on methods to standardize place names. Each course lasts two weeks and consists of lectures on principles, policies, and procedures of names standardization (largely based on BGN practices), work in the field to collect and verify local names, and office work to review results of field activities, to develop and manage names files, and to publish results. A major element is the involvement of digital processes. The twelfth course was held in Paraguay in 2000. The program has improved the abilities of participating countries to work on place names. Venezuela, for example, benefited to the extent that it created a new national authority in 1992.

Appendix J identifies how to obtain information about place names from the U.S. Library of Congress and the United Nations.

NOTES

1. Scholars and cartographers for centuries recognized the confusion caused by differing names but little was done on an international basis until about 113 years ago. The 1887 International Geographical Congress in Berlin focused on one aspect of the problem—how different writing systems produced varying spellings or graphic representations of place names that could not be universally understood. Delegates initiated efforts as least to recognize the problem and to recommend ways to convert names spelled in one system to the spelling in another system. While this was a challenge at that time (and still is), the major problem concerning place names was to create national bodies to deal with names in their own territories.

2. Cited by *the First Report of the United States Board on Geographic Names, 1890–1891* (Washington, D.C.: Government Printing Office, 1892).

3. The foreword of the sixth BGN report notes the publication included all decisions made from the board's earliest actions, while incorporating any authorized modifications. The report has three parts. Part I has statements about the nature of names, problems associated with treating names, experiences with names in the United States, and policies for working with names in foreign areas. Part II covers the board's operating principles, policies, and procedures and a brief history of the organization. Part III has transliteration tables, a pronunciation key, abbreviations associated with names, and an alphabetical list of approximately 29,000 place names acted on since 1890. It includes foreign names published in a separate volume in 1932.

4. This figure does not represent the number of individual or separate features. Instead, it covers the number of names subject to research and processing to assure correct placement and spelling for maps and charts. Nevertheless, the total number of features was vast.

5. The author was hired by DMA in 1973 to succeed Dr. Burrill and remained with that agency until his retirement in 1993. As of that year, Roger Payne serves as the board's executive secretary as well as the executive secretary of its Domestic Names Committee. Randall Flynn, of NIMA, is the board's executive secretary for foreign names. These positions are staffed with the concurrence of BGN member agencies. Until 1993, the BGN executive secretary has been closely involved with foreign place names. Over the years, the fact that two persons held three positions with the title "executive secretary" led to certain confusion, especially among foreign organizations who needed to correspond with appropriate BGN officers.

6. Marcel Aurousseau, secretary of the PCGN for many years, was a prolific writer on geographic names. The academic veracity of his works well supported the practical mission of the PCGN. His publication "Geographical Names for International Use," while only several pages long, was printed as part of the *Proceedings and Transactions of the Fifth International Congress of Onomastic Sciences* in 1958. Its bibliography contains twenty-eight references to relevant documents. The BGN Domestic Names Committee at the U.S. Geological Survey has a copy. Another text is his *Rendering of Geographical Names*, the Hutchinson University Library, 1957, 178–202 Great Portland Street, London, United Kingdom.

9

The United Nations
Joins the Effort

THE UNITED NATIONS LOOKS AT PLACE NAMES

Representatives from several countries having experiences in place names and cartography persuaded the United Nations in 1955 to study issues related to erroneous or inconsistent place names and develop appropriate corrective policies. Subsequently, under the auspices of the U.N. Economic and Social Council (ECOSOC), a group of experts met several times.[1] Following their recommendations, the United Nations convened the international Conference on the Standardization of Geographical Names in Geneva, Switzerland, in 1967.

Delegates from fifty-five countries attended, as did representatives of such bodies as the International Hydrographic Organization and the International Cartographic Association, whose work also was closely tied to accurate place names. They considered several topics such as the effectiveness of existing national standardization programs, terms needed to enhance toponymic work, requirements for national and international programs to transfer names from one writing system to another, and recommendations to continue international collaboration along specified lines.[2] The conference established committees to deal with agenda subjects, and participants worked in committees according to their experience and skills in relevant fields.

The conference produced twenty resolutions that related to a range of topics. A number of them covered procedures to develop and apply romanization systems.[3] Perhaps the most important resolution of the conference—and according to some, the single most significant U.N. resolution on place names—was resolution no. 4, National Standardization. This item provided a formula that, if fully applied, could enable any country to create and implement a program to satisfy virtually every requirement of names standardization.[4] The result would then conform to a primary U.N. principle: international standardization is based on national standardization.

Another resolution called for a second U.N. names conference, which took place in London in 1972. Its agenda was similar to that of the first conference. In the interval, nations increased their efforts to standardize names. The London conference was attended by representatives from fifty-seven countries, most of whom were at the earlier meeting. As before, some participants also came from several international bodies. Subsequent conferences convened in 1977, 1982, 1987, 1992, and 1998. As noted in chapter 14, "The Cold War," political positions related to that period affected some U.N. deliberations.

The seventh conference, scheduled for 1997, was delayed one year because of other political issues. It was to take place in Iran in that year, but the United Nations canceled it because Iran failed to include Israel in its formal invitation list. Such action (i.e., to exclude a U.N. member from a standard event) contradicts U.N. policy. At the same time, the United States said it would not attend. The conference was deferred and was held January 1998 at U.N. headquarters in New York. That meeting brought some 187 participants and observers from sixty-four countries (including Israel) and thirteen organizations involved with place names or related topics.

United Nations names conferences have produced a total of 159 resolutions and hundreds of technical reports that have helped to identify and resolve many issues. But as noted on pages 87–88, some questions have arisen as to certain aspects of U.N. programs.

U.N. GROUP OF EXPERTS ON GEOGRAPHICAL NAMES

One action of the initial U.N. conference in 1967 was to create the U.N. Permanent Committee on Geographical Names, later called the U.N. Group of Experts on Geographical Names (UNGEGN). Its purpose is to implement U.N. and related names programs. The UNGEGN is composed not of national delegates but of experts representing divisions that consist of individual countries and groups of countries having common languages or being in a stipulated region. In some cases, nations can belong to more than one division. UNGEGN working groups are created to address specific tasks and report to the UNGEGN and to conferences as well. Some recent subjects are romanization systems, terminology (associated with the field of geographic names), toponymic data files, and training. The reports include recommendations for actions that, upon further review by committees formed at U.N. conferences, may be adopted as formal U.N. resolutions. Divisions are called upon to encourage their member nations to implement U.N. recommendations as fully as possible.

Originally, the UNGEGN included fourteen divisions, but structural changes occurred that reflected dynamic linguistic and political conditions. The fifth U.N. conference in 1987 acted to form two new divisions. The Asia-Southwest Division (other than Arabic) was divided into two bodies: a new one with the same name and a second new unit, the East Mediterranean Division (other than Arabic). The latter division originally consisted of only one country, Israel, but Cyprus was added several years later.[5] The 1987 conference also created a new Celtic division to represent the cultural heritage shared by Ireland and parts of other regions, including Wales and sections of France. The Celtic ties are based principally on a common linguistic background in the cited areas, which has influenced current place names.

At the 1998 U.N. conference, a new group, the French-Speaking Division, was formed. The pertinent resolution says that since the French-speaking countries face a common set of issues from both a toponymic and cultural point of view, their work on names would be facilitated by being able to work as a division within the UNGEGN. In addition to France and Canada, member countries can include any other nation in Africa, the Caribbean, and Southeast Asia, where French is spoken by either a majority or a minority of the population. Some U.N. delegates may have felt such a division was complicated and too broad, while at the same time reflecting a pro-French language perspective. But such reservations are largely invalid, since UNGEGN procedures permit nations to join any division (or divisions). A case in point is the Dutch- and German-Speaking Division, which includes Austria, Belgium, Germany, the Netherlands, South Africa, Suriname, and Switzerland because they have common language backgrounds. Like the new French-Speaking Division, members represent a very broad distributional pattern.

In 1999, the UNGEGN divisions and their associated countries were the following:

Africa Central (*Angola, Burundi, Cameroon, Central African Republic, Chad, Congo, Democratic Republic of the Congo, Equatorial Guinea, Gabon, Rwanda, São Tomé, Príncipe*)

Africa East (*Botswana, Ethiopia, Kenya, Lesotho, Madagascar, Malawi, Mauritius, Mozambique, Seychelles, Swaziland, Uganda, United Republic of Tanzania, Zambia, Zimbabwe*)

Africa South (*Botswana, Lesotho, Malawi, Mozambique, Namibia, South Africa, Swaziland, Zambia, Zimbabwe*)

Africa West (*Benin, Burkina Faso, Cape Verde, Côte d'Ivoire, Gambia, Ghana, Guinea, Guinea-Bissau, Liberia, Mauritania, Niger, Nigeria, Senegal, Sierra Leone, Togo*)

Arabic (*Algeria, Egypt, Iraq, Jordan, Kuwait, Lebanon, Libyan Arab Jamahiriya, Morocco, Oman, Saudi Arabia, Sudan, Syrian Arab Republic, United Arab Emirates, Yemen*)

Asia East (other than China) (*Democratic People's Republic of Korea, Japan, Republic of Korea*)

Asia Southeast and Pacific Southwest (*American Samoa, Australia, Brunei Darussalam, Cambodia, Commonwealth of the Northern Mariana Islands, Indonesia, Lao People's Democratic Republic, Malaysia, Myanmar, New Zealand, Nauru, Papua New Guinea, Philippines, Singapore, Thailand, Vietnam*)

Asia Southwest (other than Arabic) (*Afghanistan, Azerbaijan, Cyprus, Islamic Republic of Iran, Pakistan, Turkmenistan, Turkey, Yugoslavia*)[6]

Baltic (*Estonia, Latvia, Lithuania*)

Celtic (*France, Ireland*)

China (*China*)

Dutch- and German-speaking (*Austria, Belgium, Germany, Netherlands, South Africa, Suriname, Switzerland*)

East Central and Southeast Europe (*Albania, Bulgaria, Croatia, Cyprus, Czech Republic, Greece, Hungary, Poland, Slovakia, Slovenia, The Former Yugoslavian Republic of Macedonia, Turkey, Ukraine, Yugoslavia*)[7]

Eastern Europe, Northern and Central Asia (*Armenia, Azerbaijan, Belarus, Bulgaria, Kyrgyzstan, Russian Federation, Ukraine, Uzbekistan*)

East Mediterranean (other than Arabic) (*Cyprus, Israel*)

French-speaking (*Belgium, Benin, Cameroon, Canada, Côte d'Ivoire, France, Lao People's Democratic Republic, Luxembourg, Mali, Monaco, Romania, Switzerland*)

India (*Bangladesh, India, Pakistan*)

Latin America (*Argentina, Bolivia, Brazil, Chile, Colombia, Costa Rica, Cuba, Dominican Republic, Ecuador, El Salvador, Guatemala, Haiti, Honduras, Mexico, Nicaragua, Panama, Paraguay, Venezuela*)

Norden (*Denmark, Finland, Iceland, Norway, Sweden*)

Romano-Hellenic (*Belgium, Canada, Cyprus, France, Greece, Holy See, Italy, Luxembourg, Moldova, Monaco, Portugal, Romania, Spain, Switzerland, Turkey*)

United Kingdom (*Guyana, South Africa, United Kingdom of Great Britain and Northern Ireland, New Zealand*)

United States of America–Canada (*Canada, United States*)

Each country of the world belongs to one or more divisions, but not all member nations of a division are able to attend every UNGEGN session, which are scheduled immediately before and after each U.N. conference and twice in between. Such partial attendance tends to frustrate the mission of the Group of Experts to collaborate with and speak for all its divisional members. In addition, many of the member countries do not have individuals who truly are experts in place names. The U.N. staff in New York supports the UNGEGN with a newsletter and publicity efforts to describe the benefits of national and international names standardization. However, the overall burden on the staff is daunting, and present U.N. budget reviews will likely reduce such support. If U.N. or other international names programs are to succeed, individual countries and organizations involved directly with names will have to provide greater assistance.

HOW U.N. PROGRAMS WORK

While the United Nations and its participating nations deserve credit for developing an international system designed to standardize place names, it is only fair to admit that shortcomings still remain. One major question is whether the U.N. directive to have only "one standard name for each named feature" remains valid. The value of this principle is that if each place has a single official name, confusion as to the nature and location of features would be eliminated. This principle was enunciated at the first conference in Geneva, Switzerland, where, however, the existence of different names for individual features seemed to contradict the goal. Geneva itself is a case in point. Geneva is the conventional English name, but in Switzerland it also is Genf (German), Genève (French), Ginevra (Italian), and Genevra (Romansh). Although the different spellings have principal status in the areas where each of the four languages predominates, in some instances two or more versions may be used. Since such multilinguistic situations exist in a number of countries, an objective position could call for a substantial modification of the U.N. view seeking one name for each named feature.

National and divisional representatives attending U.N. conferences submit papers on agenda topics. Over the years, the number of papers has totaled perhaps several hundred. Many are important to the agenda and are of interest to participants. Too often, however, the papers are overly detailed, concern matters not directly related to agendas, or describe only national programs and thus provide little information that can benefit other countries. Nonetheless, all are processed by U.N. staff for circulation to participants during meetings and later as official documents. Although submitting such reports to U.N. meetings is a national prerogative, it is a logical U.N. policy to seek papers that are practical and succinct and can be useful to a broader audience, namely, nations seeking guidance to create and manage names programs.

Resolutions are intended to be the principal vehicle to establish U.N. programs that lead to adequate standardization efforts at all levels. However, some resolutions are impractical or with time are overtaken by events and no longer serve a useful purpose. A conservative estimate is that today perhaps about a third of all resolutions are no longer important. Such resolutions illustrate one of the weaknesses of the U.N. system: the procedures for proposing resolutions and periodically judging their validity seem inadequate. A query to member states about ten years ago concerning the value of resolutions brought a wide range of responses and inconclusive results. Noting the difficulty of analyzing resolutions in this fashion, the sixth conference in 1992 requested action to assure "the classification of conference resolutions according to their relevance." To date there has been no response. However, Canada has submitted a publication grouping resolutions according to their subject matter.[8] Although this document is most useful, it also demonstrates the need for a policy concerning the submission and approval of resolutions.

Another item of concern is that so few experts in the field of names participate in U.N. meetings. As noted earlier, some nations active in U.N. programs have no names agencies or have organizations that, while having stated responsibilities for place names, may not meet basic standards as defined. The qualifications of the representatives of such nations are therefore uncertain, yet some of them are very active in most phases of U.N. agendas. In addition, because all nations cannot afford to attend meetings, some participants are members of national diplomatic missions located at the sites of meetings and have little if any specialized knowledge of the subjects. Further, too often national delegates will attend only one conference and play no active role afterward. Votes taken on issues (i.e., to approve a recommended resolution) may be cast by participants who have little if any experience with or knowledge of the topic. The end result is that most action on a continuous basis is taken by representatives of fewer than twenty-five countries that are able to provide travel support on a regular basis.

Interestingly, while the Netherlands, the United Kingdom, and several other nations have made substantial contributions to U.N. programs, have impressive national mapping programs, and have collaborated with other nations on a bilateral basis, they have no agency to standardize their own national names in a manner that appears to conform with procedures endorsed by the United Nations. Having such capable participants could seem to argue against the U.N. policy to encourage nations to establish names agencies. At the same time, it may appear illogical for such states to encourage others to create national standardization authorities.

The United Nations was able to develop many useful names programs, but the Cold War affected certain categories of deliberations between various countries. Chapter 14, "The Cold War," describes some cases in which national positions prevented full collaboration. With the demise of the Soviet system, nations are now able to meet freely and to collaborate on programs of mutual benefit. But newly independent nations such as Ukraine, Belarus, Turkmenistan, and Khazakstan are so occupied with the rehabilitation of their economies that they can apply only limited resources to names programs. Compared with conditions hardly a decade past, the current atmosphere of global collaboration is to be applauded. Given the political/cultural dynamics of our world society, however, it is not possible to guarantee complete freedom of communications in the future.

Another area requiring greater attention is the need to build stronger relationships between cartographic and names agencies at the national level. Virtually all nations with such ties have experienced many benefits and in fact, U.N. resolutions recommend such collaboration. Such joint efforts, of course, imply the existence of competent cartographic and toponymic bodies. But even where such organizations exist, success is limited. There may be no national directive requiring the agencies to work together, or the degree of stipulated cooperation may be insufficient. Another way to promote names as essential elements of mapping programs is to encourage international cartographic conferences—both official and professional—to aggressively promote programs to create national names offices and to work more closely with corresponding U.N. bodies.

A written report submitted by the UNGEGN chairman, a South African, to the twentieth UNGEGN session in January 2000 noted that despite benefits U.N. programs have brought to a number of nations, they have failed to reach many developing countries. He stated that only a small number of countries in Africa, south of the Sahara, had developed even rudimentary programs, and he also said that only relatively few nations south of the equator had names agencies. While not recommending new procedures, he concluded that the situation poses one of the major challenges facing the U.N. in the twenty-first century.[9]

FUTURE U.N. EFFORTS

Although the U.N. mission to promote names standardization is indeed worthy and has led to benefits at local, national, and international levels, U.N. programs need to stay focused on significant issues and develop actions accordingly. There is concern that some activities are not practical enough and may reflect certain personal rather than national perspectives. As noted above, national mapping and place name agencies need to work together. Further, the growing awareness of minority groups in many countries is producing an increasing view to give their languages official status and to recognize relevant place names. Such measures will have a greater impact on U.N. resources already burdened with existing place name programs. As the world looks at requirements for accurate place names, some aspects of U.N. work may require revision. In any case, it is essential that the needs of the average citizen be the primary consideration and that communications at the local, regional, and international levels be served.

The lack of adequate progress in standardizing place names has been generally recognized and the 1992 conference took note of recommendations to upgrade U.N. work. One paper addressed a need for measures helping to develop greater efficiency. It contained a recommendation to encourage collaboration within the UNGEGN divisions so member nations having effective programs could assist other nations in said divisions to create their own organizations. A resolution the conference passed also modified the UNGEGN's statement of aims and functions so that body could operate more effectively. Another resolution requested the creation of a publicity committee to publish documents defining the need for international and national standardization and describing U.N. efforts in that area. A resulting flier was circulated at the twentieth UNGEGN session, and it was agreed to disseminate it to national and international bodies involved in cartography, geography, and place names. Informing such organizations about U.N. programs is expected to help them comprehend the benefits of standardization and collaborate accordingly.

Individuals regularly involved with U.N. programs have expressed views concerning a need to review and improve them. Such views were repeated by delegates to the twentieth UNGEGN session in January 2000. But perhaps it would be even more logical for the United Nations to appoint a separate group to study the issues and make suggestions. The group could consist of two categories of people. One category could be those long active in the practical and applied aspects of place names, such as those involved with cartographic design and production, who could evaluate the background and purpose of the U.N. names programs. The second body could be those not directly associated with current U.N. names programs but representing private citizens and user groups who could perhaps be more objective as to issues on the agenda. Their joint recommendations could be valuable.

NOTES

1. The U.N. Economic and Social Council (ECOSOC) supports U.N. names programs in many ways. Meetings occur under U.N. auspices and in its facilities or those of host nations. Prior to sessions, it disseminates announcements and other relevant communications to all nations. At meetings, it provides support services such as interpretations of all proceedings in the six U.N. languages—Arabic, Chinese, English, French, Russian, and Spanish—and circulates copies of all papers submitted before or during meetings. The overall costs are substantial.

2. *United Nations Conference on the Standardization of Geographical Names,* September 1967, vol. 1, "Report of the Conference."

3. Chapter 12, "If Your Language Is English, How Do You Write or Pronounce Arabic or Russian Names?" discusses romanization.

4. Chapter 1 provides a definition of "standardize" as related to place names.

5. The formation of the new East Mediterranean Division (other than Arabic) in 1987 resulted from action at the 1984 UNGEGN session when a representative from Iran requested that Israel be removed from the Asia Southwest Division (other than Arabic). The question was taken up by the 1987 conference, and the new unit was created with Israel as its single member. Subsequent action added Cyprus to the division.

6. The United States does not recognize a nation with the name Yugoslavia. On the other hand, most countries and the United Nations accept it.

7. The cited names might be justified because they reflect languages that have official status in the mentioned countries. Nevertheless, having two or more official names for a specific place tends to contradict the stated U.N. policy.

8. Seventh U.N. Conference on the Standardization of Geographical Names, New York, 13–21 January 1998, *Measures Taken and Proposed to Implement United Nations Resolutions on the Standardization of Geographical Names,* E/CONF.9/CRP.1, 26 November 1997.

9. Peter E. Raper, *Report of the Chairman,* Information Paper no. 7, U.N. Group of Experts on Geographical Names, Twentieth Session, New York, 17–28 January 2000.

10

A Need for Place Names
Where People Don't Live

The world demands names for many places even when nobody lives there. From earliest times, maps of unexplored or uninhabited areas often carried names for places only thought to exist. Further, explorers knew it was important to give names to imagined or actual rivers, mountains, seas, and other features in records or on maps if their travels or theories about the nature and location of features were to appear valid. Many ancient maps show sea bottom and planetary nomenclature. Today, scientific research reveals ever more features on sea bottoms, in Antarctica, on the moon, and on other planets. For a variety of reasons, such features require names.

UNDERSEA FEATURES

Early maps of seas generally identified only islands and continental outlines. Occasionally, a map would show known or imagined underwater features if mariners believed them hazardous to navigation and to be avoided. For example, off the coast of the northeastern United States, a relatively shallow sandbank presented a threat to early navigation and was named Georges Bank. It remains a threat. The need to identify potential dangers was matched by a dual requirement resulting from scientific ocean-bottom surveys that revealed numerous features. First, it was natural for their discoverers to want to name them. Second, as records were compiled, maps were drafted, and additional research was pursued, naming such features was the only practical way to assure the distribution of information about their nature and location to the scientific community.

The United States had long experience with names of undersea features, but it was not until 1963 that BGN created a special committee to deal with them. The resulting Advisory Committee on Undersea Features (ACUF) brought together people experienced in oceanography, hydrography, and cartography to develop a program. The members function as board representatives, but by law they belong to various U.S. agencies responsible for producing nautical charts and portraying and analyzing sea-bottom features for scientific purposes. They are appointed by the secretary of the interior in keeping with the board's charter, and they serve for unlimited periods. There may be six members at any given time. With support from their agencies and the board, they have created an impressive record of accomplishments. They developed terms defining undersea feature names and procedures for naming them. The committee published four editions of gazetteers (the last in 1990), each of which includes a statement about committee operating procedures, as well as a glossary, and contains lists of names and associated locational data. This publication contains about 10,000 names. Appendix D shows several entries in the 1990 edition of the gazetteer. The gazetteers have forms that name proposers can complete and submit to the committee for consideration. The publications are available to the public. At present, they are being converted to a digital format, but efforts are being directed to continue with user-friendly printed editions.

The committee meets as required under the leadership of an elected chair or vice chair. Staff is provided by the National Imagery and Mapping Agency. The committee reports on its programs, and recommended names are submitted to the board for discussion, approval, or further review if necessary.

Other nations have programs to name undersea features, and ACUF cooperates with them to assure maximum validity of names to identify terrain elements. In addition, the committee collaborates with the International Hydrographic Organization (IHO), headquartered in Monaco, which since 1921 has worked with many nations to produce a chart of world water bodies, the General Bathymetric Chart of the Oceans. The primary purpose of the IHO is to ensure that all maritime nations have access to a chart whose common specifications would provide maximum navigational safety. In 1947, IHO members adopted a program to use standardized spellings of names of undersea features.

While having common goals, the missions of ACUF and IHO differ. The committee provides names of undersea features on cartographic products at virtually all scales to meet U.S. navigational and scientific requirements. About fifteen years ago, the committee added to its library ten terms for small features revealed by sophisticated new surveys able to show detail at ever-increasing scales. The IHO, on the other hand, has traditionally compiled charts at a scale of 1:10,000,000. Thus ACUF has produced many more names and for many years saw no requirement to inform the IHO. At the same time, the organization added a series of regional charts at a scale of 1:1,000,000 and felt justified in asking the committee to coordinate all its decisions. The IHO also has sought to establish itself as the international authority on undersea names. BGN has not accepted this action, since it sees its programs as fully able to meet the requirements of the broadest possible user group. Nevertheless, ACUF has agreed to collaborate by exchanging proposals for new names. Generally both bodies accept most proposals, but the committee sees no necessity to endorse all names approved by the international organization.

ANTARCTICAN FEATURES

In 1838, the United States was the first nation to send an expedition to explore and map Antarctica. Other nations have developed exploration and research programs. In 1943, the United States expanded its mapping of Antarctica in support of its involvement in World War II. The application of names to features vividly revealed by aerial photography began and continued after the war. The growing scientific exploration of the continent also brought a corresponding increase in place names. When re-created in 1947, the board formed the Advisory Committee on Antarctic Names (ACAN) with members representing federal agencies active in various aspects of research and mapping of Antarctica. In accordance with the board's charter, the secretary of the interior appoints members for indefinite terms. The U.S. agencies represented have included the U.S. Geological Survey, the National Science Foundation, the National Archives, and the Library of Congress. There are about six members at present, and meetings also benefit from the participation of former members and staff. Committee records show that world-renowned explorers and scientists, most of whom had spent time in Antarctica, participated in ACAN affairs over the years as members or consultants. The committee developed guidelines to assure that its naming procedures met the highest standards. Most names honor U.S. scientists and other researchers who have spent time on the continent or have otherwise engaged in significant work on topics related to Antarctica. The committee also accepts names proposed by other national authorities, provided such names meet normal standards. Predecessors of the National Imagery and Mapping Agency provided technical backing to the committee until 1995, when support was transferred to the U.S. Geological Survey, which has principal responsibility for mapping Antarctica.

Following the BGN example of publishing gazetteers, ACAN compiled a gazetteer in 1956 that carried 3,400 names. The fifth edition published in

1995 has 12,710 names and lists a number of additional or variant names (in italics) that represent spellings in other languages, misspelled names, or incorrectly applied names. The purpose of printing variant names is to assure that the gazetteer has a maximum locational value for persons who may use sources with different but possibly valid names. Unlike other gazetteers produced by the board, the ACAN publication describes each named feature as well as the origin of the name. This information is in addition to the location given in degrees and minutes of latitude and longitude. Appendix D gives an example of a name and associated information.

For many years, the committee was among only a few national agencies dealing with names on that continent. It has worked closely with the British Antarctic Place-names Committee and with similar bodies in Australia and Japan. As research in Antarctica increases, a growing number of countries have created staffs to work with names in that continent. One of the factors that complicates naming processes is that some nations claim areas and therefore are reluctant to share naming activities in such territories. An additional problem is that Chile and Argentina have claims to areas that partially overlap each other. The United States has no territorial claims and does not recognize those of any other countries, including Australia, France, New Zealand, and the United Kingdom.

The creation of the international Antarctic Treaty Organization (ATO) resulted in additional countries developing programs to survey and map the continent. Most of them have formed national names agencies to deal with nomenclature, and the United States has collaborated with them as well. ATO created the Scientific Committee on Antarctic Research (SCAR), which in turn formed a group to coordinate all national efforts to name features. Because over the years ACAN has developed such methodical procedures and has made all its information totally available, the United States has stated its preference not to subject its programs to external control. Instead, it will contribute its findings to SCAR for further distribution as that body desires.

SCAR attempted to have its programs endorsed as the international standard by the United Nations. A resolution at the 1992 U.N. Conference on Geographical Names agreed to take note of SCAR procedures and to establish a liaison with the organization. The relationship is for communication purposes only, as the United Nations has no function to approve place names.

EXTRATERRESTRIAL FEATURES

In the 1970s, the U.S. government was involved in an ambitious lunar mapping program that required names for the many features revealed by sophisticated imagery processes. Before this, the International Astronomical Union (IAU) named lunar and other planetary features, mainly for astronomers. By the mid-1970s, it had named about 125 craters, mountains, or other features. However, the work was unable to keep up with the growing lunar mapping program. In 1975, the secretary of the interior approved the creation of the BGN Advisory Committee on Extraterrestrial Features (ACEF) to provide names for U.S. maps. The new body collaborated with the IAU to assure adequate progress. One item questioned by the new advisory committee was the IAU use of Latin generic terms, a policy based on the assumption that the global scientific community would prefer a common language for names. Some examples are *mons* instead of mountain, *mare* instead of sea, *chasma* instead of chasm, and *fossa* instead of ridge. ACEF felt this practice was not valid because most countries would translate Latin words into their own languages and countries not using the Roman alphabet could not apply Latin terms. Although the committee recommended that the IAU approve English terms as circumstances warranted, no action resulted.

Over the next several years, the ACEF met periodically to examine new planetary images and applied names to some 100 newly detected features. By the early 1980s, however, the U.S. lunar mapping program declined and the IAU became more active. With time, ACEF ceased its operations. The scope of U.S. space programs has increased again, and since U.S. official maps may need more feature names than IAU can provide, the committee may become active again.

11

The Terminology of Names

WHY IS TERMINOLOGY NEEDED?

Like experts engaged in research in other fields, people working with place names have created a set of terms and definitions. The resulting vocabulary helps ensure an exchange of information with minimum confusion. This chapter discusses the requirement for the terminology of place names and give examples of common terms. It also lists major terms and definitions associated with a broad range of uses. Some of them also appear in earlier chapters on the origins and characteristics of place names. It also mentions two significant documents the United Nations has issued on this topic.[1]

As people began working on place names to meet practical requirements, they realized that procedures had to be developed to ensure uniform treatment. It was essential that the nature of named places be clearly understood when names were collected, examined for accuracy, changed when replaced by new nomenclature, corrected to ensure consistency, filed as official records, distributed in gazetteers, or applied to maps.[2] Further, the internal structure or language of names had to be defined. Some terminology is naturally related to the field of linguistics because much of the language associated with names relates to that branch of knowledge. Because experts in place names normally have a background in geography, cartography, and history and also have a command of two or more languages, the terminology of these fields also plays a role.[3] Owing to the multitude of languages and writing systems, standards of terminology vary considerably.

IS IT A PLACE NAME, A GEOGRAPHIC NAME, A GEOGRAPHICAL NAME, OR A TOPONYM?

The term "place name" has synonyms and variations. Agencies officially involved with place names commonly use the expression "geographic name" or "geographical name." There is the U.S. Board on *Geographic* Names, but the United Nations sponsors conferences on the standardization of *geographical* names and sessions of the U.N. Group of Experts on *Geographical* Names.

Another word meaning place name is "toponym." This word is derived from two Greek words, *topo,* which means a shape or form, and *nymy,* which relates to the word "name." A related word is "toponymy," which is more or less a scholarly term related to the study of place names. It can also be used as an umbrella term for the general field of place names. The terms "place name," "placename," or "place-name" are commonly used. Yet another word, "onomastics," means the study of all kinds of names, including place names.

As noted in chapter 1, the United Nations prefers the term "geographical name" and has provided a definition of it and of related terms that most nations accept.[4] Nevertheless, there are some differences in U.N. definitions as well as those associated with national practices. The U.N. glossary on names has this English-language entry: "Geographical name is a name applied to a geographical feature." (The glossary is printed in the six official U.N. languages: Arabic, Chinese, English, French, Spanish, and Russian.) This definition introduces the question, What is a geographical feature? The English-language definition in the glossary is, "A portion of the surface of the earth that has a recognizable identity." Usage of the term by the U.S. Board on Geographic Names includes both natural features and certain types of cultural features such as states, counties, cities, or other visible and nonvisible entities (including undersea features, visible mainly by means of electronic detection devices). Thus a geographic (or geographical) feature has a concept beyond the U.N. definition.

Many countries prefer the word "toponym," which, by general consensus, has the same meaning. But the same U.N. source defines a toponym in English as "a name applied to a topographic feature." Although a topographic map certainly depicts such entities, cannot a name applied to a city or forest also be a toponym? The word nevertheless has gained increasing acceptance as the most appropriate single term associated with the field and is preferred particularly by those whose work on place names has an academic or linguistic perspective. The Spanish definition of a toponym is slightly different. The primary definition translated into English is, "A noun given to designate a topographic feature." A second definition says a toponym can be a geographic name. Further, the Spanish definition of toponymy notes the term is related to a toponym as appearing on a topographic map (i.e., a map showing topographical features). Seemingly, a toponym would not be a feature name on another kind of map, such as an aeronautical chart that may not show terrain features in detail.

Thus there are variations in the terminology associated with place names. Such inconsistencies are perhaps not unexpected, given that different definitions can be part of languages. The different interpretations of terms, of course, inject a degree of nonstandardization in both national and international programs. Similarly, papers submitted to U.N. meetings as well as comments by participants can use words or phrases in differing ways. One of the challenges to the field is to develop universally consistent terms and definitions.

SPECIFIC AND GENERIC TERMS

Depending on national languages, place names may consist of differing kinds of words. In the United States and certain other countries, the formal name of a river or other natural feature usually has two words with distinct attributes. The names Hudson River and Rocky Mountains provide examples. Hudson is both a noun (as the name of an early explorer it is a proper noun) and a "specific" term. "Rocky" is an adjective that describes the nature of the terrain's surface but is also a specific term. The noun river is a "generic" term because it can relate to other similar features. "Mountains" also is a noun and a generic term for the same reason. Virtually all named natural features in the United States have two such parts to ensure adequate identification of features. In most cases, the specific term is the first part of a name, although some names have the generic as the initial word. Thus Lake Erie has the sequence reversed, but it is part of a body called the Great Lakes, whose sequence is of the common order. In any case, the use of specifics and generics has a corollary to family names. The specific is the first name and the generic is the last name.

Atlases of the United States and maps of other countries may also give a generic to each name, even though the local practice might not use generics. Thus "Danube River" might appear on U.S. maps, although local maps (e.g., Austrian maps) print only the specific (i.e., Donau). The U.S. view is that place names should accurately define natural features.

The U.S. custom is also to apply generic and specific terms to maps of international features. Examples are the Atlantic Ocean and the Caribbean Sea.[5]

NAMES OF CITIES, COUNTIES, STATES, AND COUNTRIES

BGN does not apply generic terms to names of administrative elements as villages, towns, cities, counties, states, or countries because they are not "natural" features. Pittsburgh (Pennsylvania) consists only of a specific term. In names having two or more words, all are considered part of the specific term. New York City is an example of a place with three specific elements. Similarly, the state of Ohio and the United States of America have no generic terms. The official name is defined as a title and thus has no generic term.

NAMES DEFINED AS "FALSE GENERICS"

The names of some settlements can indicate that they are natural features as opposed to a populated place. Such names were adopted by towns originating near a river, lake, or other feature whose names appeared appropriate to adopt. One example is Grand Rapids, Michigan, a community that developed near a turbulent section of the Grand River. Another example is Rio de Janeiro in Brazil. In January during the early 1600s, a Portuguese ship captain noted what he thought was the mouth of a river on the coast of what is now called Brazil. He named the feature Rio de Janeiro (River of January). Later exploration showed the broad feature was a bay and not part of a river. Meantime, the name was applied to a settlement that became the principal city of that country. The inclusion of such generic terms in city names leads to the term "false generic." There are many such names in the world that, despite their history of naming, retain their nomenclature.

CONVENTIONAL NAMES

Early explorers, traders, navigators, and others traveling to foreign areas encountered places new to them that had names. A necessary practice was to record the names. Hearing languages they did not understand, and generally having no access to written sources of names, they had to write the names in their own languages based on what they heard spoken by local people. The results were less than consistent as groups coming to the same areas at various times may have communicated with people having different speech styles. Further, travelers themselves may have given different names to identical features. Nations active in trade and exploration, not to mention colonization, collected their versions of names that, for the most part, became standard elements of their vocabulary.

Speakers of English compiled a large library of such names that, when they differed from local versions, became known as "conventional" names. Several English-language conventional names are Vienna for Wien (Austria), Munich for München (Germany), Prague for Praha (the Czech Republic), Venice for Venezia (Italy), and The Hague for 's-Gravenhage (the Netherlands). Virtually all nations have conventional names of places in other countries. For example, the Italian conventional name for München in Germany is Monaco.[6] BGN produced a gazetteer in 1977 containing 780 conventional names of populated places and other features (exclusive of the United States and its dependencies). Many conventional names have since been dropped because of a board policy to use local names to the extent feasible. Nevertheless, many conventional names are still found on school maps and other products in popular use. Further, English-language literature has a vast treasury of conventional names that are still relevant today. Appendix E shows a few additional conventional names with related information.

A major U.N. goal is to eliminate conventional names and to use names only as locally applied (or romanized if required by user groups). The existence of differing writing systems complicates this goal because, at least in the case of Roman-alphabet countries, names converted to the Roman alphabet from other systems could appear very unfamiliar. The official BGN policy, however, is to use local spellings (converted as required to approved BGN Roman-alphabet versions) while keeping conventional forms for country names. On the other hand, daily briefings of high-level federal officials on international affairs may use conventional names of places to ensure that participants more readily recognize areas under discussion. Private publishers may also retain conventional names for greater recognition by users. Tourist brochures and travel timetables published in the United States almost always refer to places by their English conventional names. For example, a travel brochure of Italy will most likely refer to Venice instead of Venezia.

SHORT AND LONG FORMS OF NAMES

The names of most nations have a short form and a long form. "Afghanistan" is the BGN short form of that country's name. The long form is "Islamic State of Afghanistan." The equivalent forms for some other countries include Belarus and Republic of Belarus, Burma and Union of Burma, Italy and Italian Republic, Jordan and Hashemite Kingdom of Jordan, Luxembourg and Grand Duchy of Luxembourg, Russia and Russian Federation, Slovenia and Republic of Slovenia, Sri Lanka and Democratic Socialist Republic of Sri Lanka, and United Kingdom and United Kingdom of Great Britain and Northern Ireland. A number of countries have no long forms: Bosnia and Herzogovina, Canada, Ireland, Japan, New Zealand, and Romania. One country has one name as its short and long form: Czech Republic. Appendix B is a list of all 190 independent states of the world as of January 2000 with, as appropriate, their short and long forms as well as associated information. The list was published by the U.S. Department of State and shows names of some countries as romanized, where required, by BGN/PCGN systems.

STANDARDIZATION OF NAMES

The term "standardization" is a basic element of U.N. and national parlance but may have different meanings. The cited U.N. glossary says that standardization is "the prescription or the recommendation of a particular graphic form or forms for application to a given feature, as well as the conditions of employment of that form or forms." A practical interpretation is that standardization is an effort to ensure that all place names and their associated attributes are accurate and that there can be little if any confusion about their spellings, locations, and official status in a given country. On the other hand, some find the word implies a bureaucratic effort to make everything the same. A dictionary definition is "to cause to conform to a standard." The word "standardization" was selected by names experts and U.N. officials after extensive discussions to adopt an appropriate term. Curiously, not all national names bodies use this word. Despite variations, however, there is general agreement among names experts that the term is appropriate. The extent to which the user public understands the word is another question. Another relevant U.N. term is "standardized name." It means "a name sanctioned by an officially or legally constituted entity as the preferred name." This phrase seems to be more easily understood.

CATEGORIZATION OF PLACE NAMES

Any study of place names leads to a conclusion that features can be grouped into similar if not identical types. An additional conclusion is that names can be organized according to their origins and functions.[7] Consequently, many people working on names have developed lists of terms to define feature types and name characteristics. This activity is in many ways essential to national and international programs involving place names. The purpose is to allow names experts to communicate among themselves more effectively, but the effort can also perhaps become overly theoretical.

The *Glossary of Toponymic Terminology* cited in note 1 has several terms also contained in another publication issued by the French Institut Geographique Nationale. Below are selected terms from those and other sources that enjoy general usage. The author has entered some revisions for easier comprehension.

Acronym. Word formed from the initial letter or letters of each of the successive or major parts of a composite term. Regarding place names, one interesting example is Soweto, which is derived from South-Western Townships, the name given a series of administrative areas in South Africa.

Allonym. Each of two or more place names referring to a single topographic feature. An example is Hull and Kingston on Hull in England. One allonym may be recognized as the official name, which in this case is Hull.

Choronym. A name applied to an areal feature, that is, a plain or flat area with recognizable limits but little variation of elevation.

Endonym. Name of a feature in one of the languages occurring in that area where the feature is located. Wien, the capital of Austria, is an example and is the German name.

Epotoponym. A place name that constitutes the basis or the origin of a common noun. An example is Champagne, an area in France famous for an effervescent wine called champagne. Another term is "eponym." (Appendix A contains a list of words based on place names.)

Exonym. Name used in a specific language for a feature located outside the area where that language has official status, and differing in its form from the name in the official language or languages of the same area. These also may be called conventional names. For example, Londres is the French form for London, and Vienna is the English form for Wien in Austria.

Homonym. Each of two or more identical names for different places. An example is San José, the name of numerous settlements in Costa Rica.

Hydronym. A name applied to rivers, streams, or other flowing bodies of water. Potomac River is a hydronym.

Odonym. A name of a street or road. Examples in the United States are Main Street, Park Avenue, and Interstate 95.

Oronym. A name of a distinctive feature of measurable relief that rises above the surrounding terrain, such as a mountain or a peak. Mount McKinley in Alaska is an oronym.

Patronym. A name of any feature given to honor an individual. Washington, D.C., is named for George Washington.

DESIGNATIONS

In BGN gazetteers of domestic and foreign areas, each name is accompanied by a "designation" that describes the nature of the named feature. The board feels such supplementary information is important because it enables people to know the feature type involved, which its name may not otherwise reveal. Designations consist of a code of two to four letters and possibly one number. Many designations are mnemonic and thus can be easily understood. The board has formed literally hundreds of designations for its gazetteers that are included as appropriate in the printed versions. The definitions of designations are in the front part of each publication, for example:

PPL. Populated place, such as a village, town, or city; regardless of size or function.

STRM. Flowing body of fresh water such as a river, stream, or brook. Ocean currents would not be included.

ADM. An administrative district such as a county or state. ADM1 is a first-order entity (such as a state) and ADM2 is a second-order unit (such as a county).

MTN or *MT.* Mountain or feature with a high elevation that is distinct and rises above surrounding terrain.

Appendix F shows a section from the BGN digital *Gazetteer of Austria* that defines designations and other terms associated with selected names. The format is common to other BGN digital gazetteers of foreign countries. Appendix D shows similar information from the BGN *Gazetteer of Undersea Features*. (It also shows that the BGN practice for Antarctican names is to apply a textual description of each feature rather than a designation per se.) In most cases, the generic term is the same as that listed in the designation column, but designations are added to assure correct identification of feature types whose generic terms may not be accurate. Chapter 12, "If Your Language Is English, How Do You Write and Pronounce Arabic or Russian Names?" also discusses terminology associated with languages and writing systems.

VARIATIONS OF TERMINOLOGY

There are other similar terms, and some countries have published gazetteers with names classified accordingly. The BGN utilizes such terms as "oronym" or "hydronym" only occasionally. As to terminology in general, the United Nations has published two documents as seen in the publications cited in note 1. The glossary of 1998 contains about 230 terms, while the publication of 1987 has about 110 terms. The contents provide a broad and almost esoteric collection of terms and definitions, not all of which are consistent and some of which seem to have relatively little relationship with practical aspects of place names. An examination of U.N. papers, as well as those submitted by national members, also shows variations in meanings. Three commonly used terms and their definitions from the cited U.N. publications are listed on the following page. Terms in the U.N. document are identified as A, and terms in the UNGEGN publication are identified as B.

Although the definitions are nearly the same, the differences in terminology and in the implications they convey can pose problems, particularly when translated into the six official U.N. languages. The 1987 document was prepared by the Documentation, Reference, and Terminology Section of the U.N. Department of Conference Services. The 1998 document may have a higher U.N. status with respect to place names. A logical question is the degree to which these and other U.N. bodies coordinate

VARIATIONS IN MEANING

ALLONYM

A. One of two or three names employed in reference to a single topographic feature.

B. Each of two or more toponyms employed in reference to a single topographic feature.

EXONYM

A. A geographical name used in a certain language for a geographical entity situated outside the area in which that language has official status, and differing in its form from the name used in the official language or languages of the area where the geographical entity is situated.

B. A name used in a specific language for a geographical feature situated outside the area in which that language has official status, and differing in its form from the name used in the official language or languages of the area where the feature is situated.

HYDRONYM

A. Toponym applied to a hydrographic feature. (*Note:* The source defines a hydrographic feature as "a feature associated with the seas and oceans or the seas and oceans themselves and all their parts.")

B. Toponym applied to a hydrographic feature. (*Note:* The source defines a hydrographic feature as a "topographic feature consisting of water, or mainly associated with water but not consisting of dry land.")

their efforts when dealing with common subjects. As a matter of fact, a resolution of the seventh U.N. names conference called for U.N. offices responsible for reports on similar or identical topics to collaborate more fully. No action resulted.

The vocabulary of place names is an essential element of the subject field. While some terms and definitions may have varying interpretations from country to country, certainly there is concurrence as to the importance of a dictionary of terms. Clearly, additional work is required to develop a library of consistent nomenclature. It is not at all whimsical to call for standardization in this field.

NOTES

1. *The U.N. Glossary*, no. 330, rev. 2, *Technical Terminology Employed in the Standardization of Geographical Names* (1987) and a similar publication, *UNGEGN Working Group Glossary of Toponymic Terminology*, version 4, presented at the seventh U.N. conference in January 1998.

2. Gazetteers are lists of names and are discussed in chapter 13, "Gazetteers."

3. An expert in place names, or a toponymist with a background in these fields is normally qualified to be involved with practical aspects of names (i.e., working to ensure that correct names are obtained for application to maps and charts and for use in other publications). However, not all toponymists play a role in the production of maps and charts.

4. As noted earlier, this book prefers the term "place name" but occasionally may also refer to "geographic name."

5. Such terminology applies principally to official documents. Personal conversations or other communications may well use only one term or the other, depending on context. Thus a person living in Ocean City, Maryland, may refer to the adjacent Atlantic Ocean as the Atlantic or the ocean.

6. As explained on page 105, the technical term for such names is "exonym."

7. Chapter 1, "What Is a Place Name?" describes names according to the nature of their origins.

12

If Your Language Is English, How Do You Write or Pronounce Arabic or Russian Names?

CONVERTING NAMES FROM ONE LANGUAGE OR WRITING SYSTEM TO OTHERS

Perhaps the greatest problem in dealing with place names on an international basis is language.[1] People familiar with only one language can rarely recognize names written or spoken in other languages. This requires map makers and others producing information about foreign places to apply names that their intended users will recognize. Although different languages pose sizable problems to those who refer to maps of various countries, it is important also to understand that the writing systems used by languages also present difficulties. For example, a map of Russia printed in that country will show names written in a manner that someone familiar only with English cannot read. And vice versa. Thus any plan to spell names in areas in which differing languages are used must convert such names to written forms that intended users can understand. In any case, a program to convert place names from one language and writing system to other languages and writing systems is an essential requisite for international standardization.

A major concern to names standardization has been *romanization*—a linguistic procedure to convert the spelling of names written in non-Roman alphabet writing systems into comparable versions in the Roman alphabet. In the so-called Western world, where the Roman alphabet is common to many languages, romanization programs are thus essential to convert the spellings of place names in countries not using that alphabet.

Because of the involvement of U.S. and British enterprises in so many parts of the world, linguists of those countries developed romanization systems to be applied to a number of non-Roman alphabet languages. Other Roman-alphabet countries also have romanization processes, for example, France. As noted below, the process works in two ways, one called "transliteration" and the other called "transcription." Some non-Roman alphabet countries have developed romanization systems for external use that are replacing those created earlier by user nations. For example, a system used for decades by the United States and the United Kingdom to spell names in China has been abandoned in favor of the Pinyin system created by the People's Republic of China about forty years ago.

LANGUAGES AND WRITING SYSTEMS

The distinction between languages and writing systems has been defined earlier, but it is useful to provide further descriptions and to define some related terms. When dealing with place names in countries having differing languages, people commonly refer to *language* as the method of writing names. For example, people may say that names in the former Soviet Union were written in Russian and names in the United States are written in English. As a matter of linguistics, however, names in the former Soviet Union were written in Russian Cyrillic (at least for official purposes), which defines the *writing system* (or *script*) in that country. On the other hand, names in the United States are written in the Roman alphabet. Every country using a writing system, irrespective of the county's language or languages, applies one of three major kinds of scripts: alphabetic, syllabic, or logographic.

Alphabetic script employs individual letters (with or without diacritics or special marks) to spell words. Alphabetic scripts include Arabic, Cyrillic, Hebrew, Roman, and Thai. Countries using the Arabic alphabet include Egypt, Jordan, Morocco, Saudi Arabia, and Iraq. Countries using a Cyrillic alphabet (with certain variations) include Russia, Serbia, and Ukraine. Hebrew is used in Israel. Roman alphabets are used by virtually all countries in Western and Central Europe and the Western Hemisphere, parts of the central and southeastern Pacific area, and in various countries in central and southern Africa.

Syllabic script employs a set of written characters, each of which represents the sound of a single syllable. Examples are Amharic (official language of Ethiopia) and Japanese Katakana and Hiragana.[2] Although the Korean script is alphabetic, it is graphically and visually syllabic because it is arranged in syllable blocks.

Logographic script uses a graphic symbol or combination of symbols to represent a single word. Chinese and Japanese Kanji are written in logograms.[3]

TRANSLATION, ROMANIZATION, TRANSLITERATION, AND TRANSCRIPTION

A discussion of methods to convert names from one writing system to another requires an understanding of the words *translation, romanization, transliteration*, and *transcription*. These words are often used interchangeably (and incorrectly). For example, the following statement has two errors: "The capital of Ukraine is Kiev. This is translated from Київ, which represents the form in that country's language, identified as Cyrillic." First, Kiev is not the translation of Київ but is rather the English conventional name form. Second, Cyrillic is not, strictly speaking, a language but is a kind of writing system or script. (Further, there are differing versions of Cyrillic, including Russian, Serbo-Croatian, and Ukrainian.) Such errors as noted in the cited sentence are common. *Translation* is precisely that: some place names can be translated into English (or other languages) and the translated name can become the standard version. In Germany a forested area known as Schwarzwald may appear on English or American maps or in reference works translated as the Black Forest. Names that could be translated are seen less often because locally used forms are gaining greater international acceptance.

The writer of the erroneous sentence most likely thought that the term *translation* was the same as *transliteration*, which refers to a procedure for converting the letters of various alphabetic scripts to equivalent letters associated with other alphabetic scripts. In the case of *romanization*, the conversion changes the letters of non-Roman alphabets into Roman letters.

A correct version of the sentence cited above is: "The capital of Ukraine is commonly known as Kiev, which is the English-language conventional name. The local official spelling in Ukrainian Cyrillic is Київ." A further useful bit of information could be: "If the name in the Ukrainian Cyrillic were romanized according to the BGN/PCGN system, the result would be Kyyiv."

With regard to transliteration, in addition to the example of Kyyiv cited above, two other romanized names follow: A place in Russia written in Russian Cyrillic is Щукипо, and when transliterated by the BGN/PCGN system is Shchukino. As is the case with many names, there is no English

conventional equivalent. The capital of Egypt written in Arabic is القـــاهرة.
The BGN/PCGN transliterated form is Al Qāhirah. The English conventional name is Cairo.

Appendix G shows a sample of Russian Cyrillic names spelled also according to the official Russian romanization of 1983 (called GOST) and
according to the BGN/PCGN romanization system approved in 1947. The
similarity of the systems is clear. Appendix H shows the BGN/PCGN romanization system for Serbo-Croatian Cyrillic, which is similar but not
identical to Russian Cyrillic.

Transcription converts the sound elements of a language of a writing
system into graphic form using a conventional system of characters and
symbols. The resulting Roman-alphabet spellings may have additional
marks to help ensure that the resulting pronunciation conforms as
closely as possible to the native. Tables for transcription systems include
detailed annotations concerning the change of subject scripts to equivalent Roman-alphabet letters. For this reason, the tables can be three or
four pages long and are judged too lengthy to be included with this book.

U.S. AND U.K. ROMANIZATION SYSTEMS

BGN, in collaboration with its British counterpart, PCGN, has published
romanization systems[4] for Amharic, Arabic, Armenian, Azerbaijani, Bulgarian, Burmese, Byelorussian, Chinese Characters, Georgian, Greek, Hebrew, Japanese Kana, Kazakh, Khmer (Cambodian), Kirghiz, Korean, Lao,
Macedonian, Maldivian, Mongolian, Nepali, Pashto, Persian, Russian,
Serbo-Croatian, Thai, Turkmen, Ukrainian, and Uzbek. Most of these systems result from several decades of work by the United States and the
United Kingdom, but as of the date of the cited publication, they also reflect efforts of some of the relevant countries. Further, in the case of Greek
romanization, Greece announced its adoption of its own system about
twenty years ago, and BGN and PCGN approved it in 1996.

The cited romanization systems do not enjoy full international acceptance, since some of the affected countries feel only they have the right to
create such standards. United Nations conferences and technical sessions
address these questions, and U.S. and U.K. representatives are willing to
consider improved systems. The U.N. Group of Experts on Geographical
Names has a working group on romanization systems that is dedicated to
the development of systems that can be accepted as U.N. standards.[5] It is
interesting to note that although some non-Roman alphabet nations have
developed romanization systems, for many other non-Roman alphabet
states the BGN-PCGN systems remain the sole conversion formulas. The
field of names standardization faces a genuine challenge when requests

are registered to modify, if not totally abandon, long-established romanization systems in favor of newly devised schemes. As such new systems become official for the covered countries, Roman-alphabet countries may experience great confusion when they revise maps, atlases, globes, and reference works. The existence of such dual romanization systems has given rise to the terms "donor system" and "receiver system." A donor system is one developed by a non-Roman alphabet country, such as Russia, for use by Roman-alphabet countries. A receiver system is one developed by one or more Roman-alphabet countries, such as the United States and the United Kingdom, for application to a non-Roman alphabet country, such as Russia. Not all Roman-alphabet countries are eager to accept donor systems as soon as they attain local official status. It is possible such new systems can be changed from time to time and it is also possible that the letter combinations and diacritics may make it difficult to pronounce names.

PRONUNCIATION OF PLACE NAMES

Is it logical to expect that in addition to standardized spellings of place names, there should also be standardized pronunciations? Can an English speaker accurately pronounce a romanized version of a place name in Russia? Referring to the name Shchukino, is even the most careful pronunciation adequate? Does one say Al Qāhirah so that a citizen of that city in Egypt will recognize the reference? For that matter, names in a given Roman-alphabet language can pose problems for speakers of other Roman-alphabet languages. How does a person speaking Spanish pronounce a place name in Holland? How does a person speaking Italian pronounce a name in Poland? Atlases and dictionaries of place names give pronunciations for listed names to enable users to speak names with reasonable accuracy. The cited BGN publication on romanization systems has guidelines concerning the treatment of specified letters or graphic symbols of non-Roman alphabet names. The guidelines relate mainly to respelling such names using prescribed Roman-alphabet letters as well as diacritics and special marking in most cases, but they also may concern pronunciation. There is no overall effort, however, to produce any degree of official pronunciation. The International Phonetic Alphabet (IPA) has a variety of letters and symbols designed to replicate the sounds associated with local and official pronunciations of words (and place names). The IPA is very complex, however, and is not familiar to many people. Thus, except for documents used by experts, it cannot be applied to the printed spelling of a name to any significant degree.

Although the letters can thus be changed according to such standards, a pronunciation resembling correct local speech patterns cannot easily be assured. One assumption by official names agencies in the United States and

the United Kingdom is that a desired standard of pronouncing vowels of romanized names is generally based on the Italian practice and that the pronouncing of consonants is based on the English custom. Thus the *a* has the
sound of "ah" as in "father," the letter *e* has the sound of "ay" as in "day,"
and so on. Using the Italian standard for vowels helps assure a more standard pronunciation, which otherwise in English would have many variations. Adding certain accents or special marks to selected letters is permitted to assure more authentic pronunciation of names. When such additions
are made, instructions as to their usage are part of the romanization tables.

Attempts to develop standards of pronunciation are subject to many
problems. Even in countries having a single national language it is not unusual to find different speech habits from place to place that produce varying pronunciations of place names (and, of course, many other words).[6]
One of the difficulties in developing a standard pronunciation for some
languages is that some of them employ tones. In Vietnamese, for example,
different tonalities or pitches given to specific vowel sounds can convey
distinct meanings.[7] Furthermore, the voice, once trained from infancy to
pronounce vowels or consonants according to local speech habits, may not
be able to make sounds associated with other speech customs. For example, persons from China or Japan usually find it difficult to pronounce the
letters *r* and *l* as normally spoken by people in the United States.

As noted in chapter 8, "Countries Recognize the Problem," the Domestic Names Committee of the U.S. Board on Geographic Names attempted
to create standard pronunciations of U.S. place names but abandoned the
effort because linguists working on the project could not locate acceptable
examples. United Nations discussions on the topic of pronunciation, more-
over, have seen no resolution.

NOTES

1. Note 1 in chapter 1 defines the technical difference between "languages" and
"writing systems." The current chapter discusses procedures for converting languages and related topics, but the subject is complex and can be covered here only
in a generalized manner.

2. The *Romanization Systems and Roman-Script Spelling Conventions.* Prepared by
the U.S. Board on Geographic Names, Foreign Names Committee Staff, and published by the Defense Mapping Agency (1994). It describes the romanization system for Japanese Kana as follows: The Japanese language is written in two forms,
namely, in kanji, which are Sino-Japanese characters, and kana, which are syllabic
symbols. There are two styles of kana, namely, katakana, the "squared" form, and
hiragana, the "cursive" form.

3. Note 2 above refers to kanji.

4. To obtain a copy of *Romanization Systems and Roman-Script Spelling Conventions,* see Appendix I, "Accessing Information about Foreign and Domestic Place

Names Processed by BGN." Since publication date, some different systems have been approved.

5. Information about this program is available at <http://www.eki.ee/wgrs>. Last accessed October 29, 2000.

6. When such speech habits represent variations of a basic language, they technically are called "dialects." The common term for such speech variations, however, is "accent."

7. Certain English-language words can also have different meanings because of tonal qualities or voice stress. As an example, the word "yes" can convey more than one meaning. It has a strong and positive sense when said as a firm statement (as though an exclamation point followed). It has a conditional meaning when pronounced with a rising inflection (as though a question mark followed). Other languages have similar traits.

13

Gazetteers

A gazetteer is a list of place names of a specified area in alphabetical order with latitude and longitude for each named feature.[1] An additional and important gazetteer function is to identify the nature of features. A gazetteer may also provide other information, such as populations of cities and towns, heights of mountains, areas of lakes, economic conditions, and photographs. For persons seeking information about places cited in the press or other literature, a gazetteer can thus provide useful background.

BGN produces gazetteers of virtually all foreign countries with names listed in strict alphabetical order. The entries also indicate feature types and the series numbers of maps on which named features are located.

During World War II and the years following, U.S. military and intelligence organizations produced many maps, charts, and other documents to keep planners informed about conditions around the globe. Except for gazetteers already produced by the United Kingdom for certain areas, there were no substantial sources of names information for many foreign territories. One significant product soon after the war was the *National Intelligence Survey*, a series of publications issued by the Central Intelligence Agency. Each had nine chapters that covered military, economic, cultural, and intelligence issues, and other topics about designated countries. Because the documents necessarily referred to numerous places, one chapter was a gazetteer giving locational and related information about features mentioned in other chapters. The gazetteers, produced by BGN, were classified documents until about 1968, when they became increasingly available for use among federal agencies. In the mid-1970s, they became public-sale items.

During much of the Cold War, the Soviet Union and its allies did not circulate publications with extensive names information. The BGN staff was thus obliged to survey a broad range of documents to locate and verify names data from Iron Curtain countries. Attempts were also made to collaborate with other countries, but responses were limited. Chapter 14, "The Cold War," provides information about U.S. experiences during this time.

At present, the board is producing gazetteers of foreign areas in digital form, but its library also includes printed publications of areas yet to be digitally covered. By 1997, the gazetteers covered eighty-seven foreign countries and provided data on some 5 million names. Appendix F shows information from the digital gazetteer of Austria that is typical of gazetteers of foreign countries.

The printed gazetteers are very similar to the digital versions. Although printed gazetteers have introductions that define information about sources and content in some detail, the digital editions require users to research the Internet for such information.

Digital gazetteers have a number of benefits. Speed of acquisition of names and the ability to print them out quickly are distinct advantages. Of course, the data is "read only," so information cannot be modified by users. However, the published printed documents may represent a preferred mode of reference for many people who find turning paper pages on a desk an efficient method of research. One can also enter corrections as needed on a printed publication.

BGN gazetteers of individual U.S. states have been published by state authorities under contract with the U.S. Geological Survey according to federal specifications. Currently, there are digital gazetteers for all states, the District of Columbia, and U.S. outlying areas. USGS also publishes a BGN concise gazetteer of the United States that contains about 42,000 names. These publications contain information similar to that found in foreign gazetteers.

A number of other countries also publish gazetteers. In virtually all cases, the items cover only the publishing country. As early as 1925, however, the PCGN began a series of gazetteers of foreign areas that were part of, or of prime interest to, the British Empire. Its peak output occurred just prior to World War II, when approximately twelve gazetteers were published. American-British collaboration on names has a long history, and when the United States became involved in the war, it recognized the value of the U.K. publications. Shortly after the war, the BGN began its own series, which were quite similar to the British model. At present, the PCGN produces gazetteers of several countries but, curiously, the body has no authority to work on U.K. domestic names. The U.K. Department of Ordnance Survey has published a gazetteer of British names that appear on their maps covering the United Kingdom.

Canada also has a domestic gazetteer program. Each province and territory has its own names authority that is responsible for local names standardization. The Geographical Names Board of Canada publishes a national gazetteer as well as one for undersea features in or near what are considered Canadian waters.

As more countries create national standardization bodies, a major goal is to publish gazetteers, albeit with varying formats and contents. Gazetteers produced by Chile, Iran, and Poland list features in alphabetical order but segmented according to feature types, such as land features, water features, and settlements. Furthermore, in some countries, gazetteers may be called geographical dictionaries because they also have photographs, text, and other information about named places.

With the growing understanding that gazetteers are key elements of names standardization, there is a greater readiness to create international exchanges of names data. The United States is now collaborating with several other nations to create mutually beneficial production procedures. This effort has been especially strong in some countries once part of the former Soviet Union.[2] A major goal is to publish gazetteers with official status for both the United States and a second country. But such efforts inherit some significant problems. Where the other country has a non-Roman alphabet script, a major challenge is to apply a romanization system acceptable to both sides. Whether gazetteers will be digital or printed is a minor point at this stage. In any case, it is likely that digital data can be exchanged during the process. As collaboration on bilateral cartographic production also becomes more common, agreements about names policies will become essential.

Gazetteers are necessary for many reasons. Publishers of newspapers, textbooks, atlases, encyclopedias, and other reference works find them essential, as do city and university libraries and the press media. With new digital techniques, even individuals can access names information on a worldwide basis. Whether printed or digitally produced, gazetteers are recognized as unique sources of information about place names.[3]

NOTES

1. Joseph T. Shipley, *Dictionary of Word Origins* (New York: Philosophical Library, 1945), 165. The word "gazetteer" comes from the Italian word *gazetta*, a magpie. When coins were first produced in Italy, they were often given the names of birds. A *gazetta* was a small coin. Early newspapers published in Italy could have cost a *gazetta*. That word has since become the English terms gazette (a printed document) and gazetteer (a list of place names, as defined in this book).

2. Representatives of BGN have had several meetings with their counterparts in Estonia and Ukraine. Both countries are dealing with issues affecting their place names, and it is hoped that board experiences can help resolve any problems the countries may encounter.

3. For information about acquiring BGN gazetteers, see Appendix I, "Accessing Information about Foreign and Domestic Place Names Processed by BGN."

V

U.S. and International Names Programs during and after the Cold War

Whereas previous chapters demonstrate that names are essential to many personal, national, and international aspects of communications, the next five show how they relate to planning and implementing not only U.N. programs but also U.S. diplomatic policies, military campaigns, and intelligence operations. The time interval covers the last half century, which includes the Cold War and gives examples of how U.S. agencies worked together to represent national interests to the maximum. The role of the North Atlantic Treaty Organization is also described. Although other countries had similar experiences, this book does not give any details on such situations.

14

The Cold War

THE COLD WAR IMPACT ON U.N. NAMES PROGRAMS

Soon after the United Nations began programs in 1955 to standardize names, the effect of the Cold War became evident. Some Western nations generally provided full details about their place names, but the Soviet Union and its allies did not distribute such data. They considered it classified and did not even distribute it to their own citizens. This one-sided policy frustrated a generally accepted U.N. view that a free exchange of names data would benefit international standardization programs. The U.S. practice of circulating gazetteers of foreign countries beginning in the early 1970s tended to highlight the fact that Iron Curtain countries generally did not have such a policy. At the third U.N. Conference on Geographical Names in Greece in 1977, the U.S. delegation mounted a display of its names work, complete with copies of each foreign gazetteer. A member of an Iron Curtain country stated his displeasure at the U.S. publication of such documents. The practice ignored what he expressed were exclusive national prerogatives. The next day, in seeming reaction to the complaint, several gazetteers were missing. They were returned later after the U.S. delegates noted their absence and informed participants that the publications were intended for public display.

A more obvious aspect of communist policy was highlighted during a discussion about the value of romanization systems at the 1987 U.N. Conference in Montreal. The United States and the United Kingdom had long argued that their common romanization system for Russian Cyrillic was supported by many countries and had numerous advantages over others being proposed. Delegates from the Soviet Union, however, noted that they had a useful system. Although it actually was not useful for wide application, Soviet allies supported the Soviet view. For example, the Cuban representative stated that his nation endorsed the system, claiming it was totally compatible with Cuban language customs. The fact that Cuba had little if any requirement to romanize map products covering the Soviet Union showed the political nature of that position.

The U.N. policy allowing member nations to attend regional U.N. meetings on place names or other topics permitted another kind of Cold War maneuver. Representatives from one or two Eastern European communist countries regularly attended U.N. names sessions in Latin America. Such representation, however, produced virtually no actions germane to Latin American names programs. It did, nevertheless, permit procommunist nations to establish contacts and otherwise promote their platforms from a political point of view. With Cuba being part of Latin America, that nation's role in fostering communist positions seemed obvious.

SOVIET MAPS WITH INCORRECT NAMES

The Soviet Union did not circulate detailed maps covering its territories even to its own citizens other than those in its military or intelligence agencies. The occasional map that was distributed most likely contained inaccurate information meant to mislead any potentially hostile users. Interestingly, a newspaper published in the Estonian Soviet Socialist Republic in February 1970 reported that a map at 1:600,000 published under the control of the official Soviet mapping agency demonstrated such a practice. The *Military Engineer* of the American Society of Military Engineers subsequently reported that place names on the cited map were incorrectly located and misspelled.[1]

The magazine also reported that Soviet atlases over a period of several years showed several cities and facilities with incorrect locations. Furthermore, a 1984 issue of the *Air Force Magazine* showed that U-2 reconnaissance aircraft in the early 1960s photographed a principal Soviet missile test and launching station at a place U.S. intelligence agencies called Tyuratam. The Soviet Union denied the place existed and reported any missile activities were at a location 200 kilometers to the northeast. Although the validity of the name Tyuratam was open to doubt, without question the evidence accurately pinpointed the location of cited activities at the location the United States called by that name.[2] The Soviet denial of such a place at that location served its policy of not identifying militarily significant places.

Some time later, a Soviet tourist map of Moscow available to local citizens as well as others was shown to have numerous inaccuracies. The Central Intelligence Agency had also produced a tourist map for U.S. representatives having official business in Moscow. Based partly on satellite imagery and enhanced by ground observations, the map clearly demonstrated that official Soviet map makers had given incorrect positions of streets, buildings, and a river, and had printed some incorrect names. The CIA product eventually became very popular even among Moscow citizens. These facts about Soviet mapping and naming practices clearly related to circumstances affecting U.N. programs.

It is important to note that during the Cold War, delegates to U.N. names conferences from all countries met with and conducted business with dignity and often with levels of personal respect and professionalism. While representatives from both sides of the Iron Curtain collaborated on a range of topics, there were occasions when political orientations clouded many issues. Thus the current atmosphere of global cooperation is to be applauded. The political/cultural dynamics of our world society, however, cannot guarantee complete fulfillment of national and international names standardization in the future.

NOTES

1. "Soviet Cartographic Falsification," *Military Engineer* 410 (November–December 1970): 62.

2. Dino Brugioni, "The Tyuratam Enigma," *Air Force Magazine*, March 1984.

15

The U.S. Department of State

The Department of State (DOS) needs accurate place names for a variety of reasons to support its mission to represent and promote the interests of the United States. DOS officials and staff correspond with officials in U.S. agencies stationed in other countries. They meet with diplomats, attend bilateral or international sessions, and collaborate with foreign organizations to develop and implement programs of mutual interest. In addition, they confer with persons from various U.S. agencies to discuss national activities involving other countries or relevant diplomatic protocols. In many cases, such actions require maps or other graphics of foreign areas that must have accurate place names.

The DOS was one of the first agencies to become part of the BGN and has always played a key role, especially regarding its responsibilities for foreign names. The Office of the Geographer, now the Office of the Geographer and Global Issues (GGI), was created in 1920, when the department recognized a need for greater knowledge about the nature and location of new countries and international boundaries. Its functions also required the collection of information about the many new place names created after World War I. Since that time, members of GGI held important positions on the board.[1]

The DOS in turn depends on the GGI to provide departmental bureaus with current names. The names are included in official reports or are used to meet requests on short notice for briefings or other presentations requiring maps. In all cases, graphic or textual communications depend on current names. When representatives from other countries may be involved, a presentation that shows even one incorrect name can lead to significant diplomatic squabbles. If the name is of a country, the problems can be considerable. Cartographic products compiled by the National Imagery and Mapping Agency for various international meetings that involve DOS or other U.S. agencies also require correct names. When such special maps are produced, NIMA seeks DOS approval of country names, the absence of which could hamper U.S. efforts to serve in a useful capacity.[2]

Accurate names are obtained by the GGI in two major ways. First, that office and other DOS bureaus receive official reports from overseas stations, from other U.S. offices, and from their foreign counterparts that can contain names information. Such names are then collected and processed as internal records. Second, DOS representatives attending periodic meetings of the BGN Foreign Names Committee (FNC) have access to foreign names processed by the FNC staff. As noted earlier, DOS members of the board provide authoritative information about major names (i.e., country names) that the staff reviews and normally accepts. If the staff identifies a problem, it studies the question and presents the case for further consideration until problems are resolved. Although the DOS has authority to decide the status of country names for official U.S. purposes, FNC reports and the BGN *Foreign Names Bulletins* are the proper vehicles to identify such names. Since all U.S. departments also are generally required to use officially correct place names in any communications with foreign governments, they also rely on names representing, in effect, joint BGN and DOS decisions.

In view of its interest in foreign names as related to its policies concerning the diplomatic recognition of foreign nations, the DOS works with U.S. delegations to U.N. meetings on names. During discussions on new names or names disputed between two or more countries, department representatives advise U.S. delegates as to positions they should take.

NOTES

1. From 1986 to 1988, the State Department member, from the Office of the Geographer and Global Issues, was chairman of the board and its Foreign Names Committee at the same time. Her talents and diplomatic responsibilities justified holding the positions simultaneously.

2. Chapter 3, "Maps Say Little without Place Names," describes NIMA's cartographic production mission.

16

U.S. Intelligence Agencies

United States intelligence agencies need accurate place names to fulfill their responsibilities to obtain and process information about a range of worldwide activities and to develop appropriate programs. Principal among such organizations is the Central Intelligence Agency (CIA). As a BGN member agency, it has provided important leadership while benefiting from board activities. Among CIA functions are preparation of reports about specified places as well as maps covering areas to illustrate the nature and distribution of various phenomena. Maps also provide cartographic background for briefings given by officials of the agency and other organizations. Such briefings are normally on a daily basis but also on short notice when fast-breaking events require quick responses. Maps and related support documents thus require access to current information, including place names, that must be processed virtually on a continuing basis.

Briefings rely on BGN-approved names, but in some cases conventional or commonly known versions of such names are used to ensure that officials being briefed might more easily recognize certain places.

As noted in chapter 13, "Gazetteers," the CIA published the *National Intelligence Survey,* a document about individual foreign countries. One chapter was a gazetteer that listed all place names (and their geographic coordinates) referred to in the document.

The agendas of the BGN Foreign Names Committee (FNC) are unclassified since virtually all of that body's work relates to open documents. Further, few of the committee staff have appropriate clearances. Representatives from the National Security Agency, the Defense Intelligence

Agency, and other U.S. organizations interested in various names issues, however, have also attended meetings as observers. On one occasion, the FNC executive secretary and other select staff members met with individuals from intelligence agencies to discuss a need to identify nicknames of places in some areas. The issue was based on this situation: aircraft or boats in the Caribbean area engaging in illegal drug trafficking occasionally communicated with their agents in the United States or elsewhere as to places they expected to land or unload drug cargoes. Such communications, which could be intercepted, occasionally referred to nicknames of places. The questions were whether the FNC staff could recognize such names or their formal counterparts, or could ascertain their locations. If so, the cited places could be found and intercessionary action taken. Unfortunately, the staff had little if any information of this kind, so the matter could not be pursued. In any case, U.S. personnel engaged in drug surveillance otherwise rely on accurate names information to verify the location of places.

Other intelligence agencies also need access to correct names, whether for internal communications or for reports to other organizations. The merging of various components of several intelligence agencies with the Defense Mapping Agency in 1996 to form the National Imagery and Mapping Agency would seem to indicate that requirements for names relating to missions of the agencies might now be coordinated within NIMA. At the same time, it is evident that CIA membership on the board will continue to represent major interests of the intelligence community.

The CIA publishes a variety of documents that are available for public purchase. One useful item is *The World Factbook*, an annual publication that provides political, economic, and cultural facts for each country.[1] It also has maps of countries and world regions, a list of all place names in the book, and numerous tables defining international organizations, country codes, and other kinds of data relevant to the scope of the publication.

NOTE

1. Requests for *The World Factbook* should be directed to the Superintendent of Documents, P.O. Box 371954, Pittsburgh, PA 15250-7954.

17

The North Atlantic Treaty Organization

S oon after World War II ended, the United States and various European countries recognized that the Soviet Union and its allies had political and military objectives endangering the peace and tranquility of Western European and other nations. The new circumstances created what became known as the Cold War. In 1949, the United States, Canada, the United Kingdom, France, and other Western nations reacted by establishing the North Atlantic Treaty Organization (NATO), with the aim of promoting their mutual defense and economic cooperation. With time, the scope of NATO functions enlarged, as did its membership.

Possible plans for executing military ground, air, and sea operations required the development and application of numerous programs related to military operations. One program produced an agreement on place names, *Standardization Agreement (STANAG) 3689: Place Name Spelling on Maps and Charts.* Its purpose is to ensure that NATO maps and charts produced by member countries carried standardized or consistent place names. The policy was in keeping with principles that most Western countries had long understood, namely, that military units from different nations involved in collaborative operations had to refer to common place names on their maps and charts. Lacking such a standard could compromise even the simplest operation. NATO maps cover a wide expanse of territories in which different languages and writing systems are used locally. A major purpose of STANAG 3689 is to provide romanization systems for countries in which the Roman alphabet is not used. Reflecting U.S. leadership in cartographic and names programs, most systems are those approved by BGN and its British counterpart, the Permanent Committee on Geographical Names (PCGN). Romanization tables are included for the following languages: Arabic, Armenian, Azerbaijani, Belorussian, Bulgarian, Georgian, Hebrew, Kazak, Kirghiz, Macedonian, Persian, Russian, Serbo-Croatian, Tajik, Turkmen, Ukrainian, and Uzbek. The document also shows transliteration tables for Arabic (for Egypt, a system created by Egypt) and for Lebanese Arabic (a system produced by France), as well as a transcription table for Greek (a system developed by Greece).[1] The STANAG permits some variations of standardized names during phases of product revision, and it also allows cartographers to show different names in parentheses to satisfy local naming customs.

In the late 1980s, there was a growing movement by U.N. members to adopt other romanization systems for individual nations. Such systems were, in most cases, radically different from the BGN/PCGN practices, which, in any case, had become part of the broad English-speaking world. Prior to the fifth U.N. Conference on Geographical Names in 1987, the United States had realized that romanization systems stipulated in STANAG 3689 could be abandoned by some NATO member states that preferred different local systems. Such action would result in one set of names for NATO products being different from those officially used by some member states. Although national representatives of NATO reiterated their support for STANAG 3689, the United States informally asked the representatives to request their national delegates to the conference to not support any U.N. resolutions backing new or different systems. Despite the assumption that there should be a single national position on romanization systems, some U.N. delegates from NATO countries supported new systems. This situation highlighted the difficulties inherent in a variety of U.N. programs on names: delegates may act independently of national policies.

The end of the Cold War, as well as the birth of a number of independent nations, has opened up many new and positive avenues. The United States now enjoys close working relationships with the Baltic states, Poland, Russia, Ukraine, and other new nations.

The conditions leading to the establishment of NATO have been markedly altered, but the organization continues its function of securing mutual defense and cooperation. As of 1998, the original twelve member nations had grown to sixteen, and in 1999 the Czech Republic, Hungary, and Poland were granted membership. Strife between the province of Serbia and the nominally autonomous province within it—Kosovo—in what many call Yugoslavia brought NATO into military action to terminate Serbian attacks against the Albanian population of Kosovo.[2] Cartographic products reflecting STANAG 3689 were employed.[3]

NATO also issued a document regarding gazetteers, *Standardization Agreement (STANAG) 2213: Gazetteers.* It gives specifications that member nations should follow in the production of gazetteers either of their own territories or, through proper coordination, other areas. Generally, the names would be those on NATO designated maps and charts, but additional names can be used if they update those on source maps or if they represent valid alternative names. The specifications closely follow those defined by NIMA and its predecessor organization, the Defense Mapping Agency. Such specifications, in turn, are BGN/PCGN standards. Given the growth of national agencies responsible for place names and gazetteers, specifications for such lists of names now being defined by NATO member countries may be modified. With continuing U.S. participation in international names programs, however, it is somewhat likely that such new gazetteers will adhere to the BGN format.

NOTES

1. Although all of these systems fall under the category of "romanization," the noted terms reflect titles accompanying documents provided by the cited countries.

2. Yugoslavia is the name commonly used in the U.S. press and in other media. The U.S. policy is, however, that even though the former Socialist Federal Republic of Yugoslavia no longer exists, the name currently used by its successor government, the Federal Republic of Yugoslavia, cannot be applied. That remaining administrative territory now includes only Serbia and Montenegro (locally called Srbija and Crna Gora): the official U.S. name for that collective area is Serbia and Montenegro. Kosovo is, according to the government of Serbia and Montenegro, a part of the province of Serbia. Ethnic Albanians in Kosovo compose a majority of the population of that territory and have declared a desire to become an independent state. Early in 1999, Serbian military action was mounted to subdue Albanians. NATO forces attacked Serbian targets in a campaign to eliminate Serbia's opposition to what NATO felt was a legitimate goal of the Albanian population.

3. In May 1999, NATO aircraft bombed what was shown on a chart to be a military building in Belgrade. The structure was actually the Chinese Embassy. American planners later admitted that the attack was based on a 1992 chart that had not been updated to show that the intended target was no longer at that location. Although such misidentification did not involve a place name, the problem is pertinent. A military operation directed against a specific place must be based on accurate information about the nature and location of a target.

18

After the Cold War

The end of the Cold War brought a wide and dramatic series of changes to virtually the entire world. The most significant modification was that the former Soviet Union and its allies no longer posed an economic or a military threat to the Western world. European nations that were part of the Warsaw Pact dissolved that organization in 1991. Other countries such as China and North Korea retain their communist forms of government, but for the most part they do not have overtly hostile ambitions regarding the United States or other Western governments.

With regard to place names, the new intergovernmental relationships quickly produced new liaisons between formerly hostile nations. The sixth U.N. Conference on the Standardization of Geographical Names in New York in 1992 for the first time saw representatives from Russia, Ukraine, Belarus, and other independent countries formerly part of the Soviet Union. In some cases, delegates of these and other countries also had represented the respective Iron Curtain nations at earlier U.N. sessions.

Participants from Ukraine and Estonia developed ties with U.S. representatives at U.N. meetings that have proved long-lasting. In 1997, the executive secretary of the BGN Foreign Names Committee and one of the committee's scientific linguists were invited to Ukraine to present a course on names standardization. The event gave Ukrainian names experts valuable insights as to methods employed by the board to work not only with its domestic names but also with names in other countries. Additional topics included digital treatment of place names and romanization. The close professional relationship continues to benefit both countries.

The delegate from Estonia demonstrated a high degree of competence and soon became a close collaborator with the United States. Under his leadership and through continuing correspondence with board staff, Estonia has developed an impressive names organization that has brought benefits to that country and also to neighboring nations and others.

With the disappearance of political tensions related to the power of the Soviet Union, the United States looks forward to collaborating with former communist nations that are now independent. The new level of international communication, however, will bring some challenges to U.S. programs. An increasing number of nations with non-Roman alphabet writing systems will strive to develop their own romanization systems for international use. The United States will attempt to persuade such countries to utilize the systems BGN and its British counterpart, PCGN, have developed or to develop "donor systems" similar to BGN/PCGN systems. In any case, one certain development will be that the United States can rely on names organizations in those countries to provide information that for so many years was difficult to obtain. In this way, the board will create mutually beneficial relationships, as it has done with names agencies in Canada, the United Kingdom, and a number of other nations.

VI

Interesting and Unusual Names

Names attract the public's attention for various reasons. On the one hand, such names can be the subject of political dissension. Names can be so closely associated with people and cultures that attempts to change them—or not to recognize them—can generate local and international disputes. Other names may become well known because they are humorous, are associated with interesting kinds of features, or are seen as unacceptable for various reasons. The final two chapters of this work provide a brief survey of place names that receive publicity because of such characteristics.

19

Names in Dispute

Disputes over names can originate for a variety of reasons and can range from mild to internationally serious. As authorities try to choose a single name for a feature having two or more names, some local residents may object to the selection of a single name. Nations gaining political jurisdiction over an area may apply new names to features. People whose cultures may be reflected by existing names will dispute new ones and will restore former ones if and when conditions allow such action. Names used for a long time can be disputed by people who find them culturally offensive and who seek suitable substitutes. Names in dispute are perhaps more evident today than in previous history because publicity concerning them has become so widespread. Whether local or global, they are an inherent characteristic of the world of names.

SOME DISPUTES IN THE UNITED STATES

Mount McKinley or Denali in Alaska

A prospector in Alaska in 1896 named a high mountain after William McKinley of Ohio, who had been nominated for the presidency of the United States.[1] There was early opposition to the name but it faded with the assassination of President McKinley in 1901. Nevertheless, a number of people believed the mountain should be called Denali, the local Athabascan Indian name, which meant "the big one." Over the years, there have been many requests to adopt that name. As is the case with similar requests for a change, arguments covered a number of factors. The name "Mount McKinley" had acquired worldwide fame because it identified what later was found to be the highest peak of North America. Schoolchildren in the United States came to know that fact. Logic thus indicated that any change would run counter to a tradition. Further, although many in Alaska accepted the name Denali, some contrary evidence could not be totally ignored. Since the Athabascan people did not have a written language, there was no official record that Denali had been used. What's more, the term "Denali" could have been used by Athabascans living in many other places to describe all large mountains. In 1975, however, the state of Alaska officially requested the U.S. Board on Geographic Names to change the name. The question received wide publicity, and in 1977 the U.S. Congress stepped into the case, an action that removed the board from further involvement. One supporter of the name "Mount McKinley" was Rep. Ralph Regula of Ohio's Twelfth Congressional District, the same district McKinley represented prior to his election as president. The historical and political nature of his request to maintain the name was obvious. Subsequently, Congress worked out a compromise: the name of the peak would remain Mount McKinley, while the area below the peak would be designated Denali National Park. In Alaska, Denali is commonly used as the mountain's name.

Mount Rainier or Tacoma in the State of Washington

In the late eighteenth century, Capt. George Vancouver of the British navy explored parts of the western coastal area of North America and named several features. Seeing a prominent mountain in an area now known as the state of Washington, he named it Mount Rainier in honor of his colleague, British Admiral Peter Rainier. About a century later, a land developer began work in a town called Tacoma, also said to be the local Indian name for the cited mountain that was near the settlement. Maps issued by the Tacoma Land Company and the Pacific Northern Railroad (located in Tacoma) applied that name, Tacoma, to the mountain to publicize their commercial interests. But the town of Seattle resented the competition generated by Tacoma and worked to retain the earlier name. When the Pacific Northern Railroad later moved its headquarters to Seattle, arguments favoring the name "Tacoma" lost some status. The debate, however, continued. During the first quarter of the twentieth century, Congress again reviewed the case, but by 1924 it had taken no further action and the name "Mount Rainier" survived. Although the issue now seems settled, local efforts to change the name continue.[2]

Cape Canaveral or Cape Kennedy in Florida

The assassination of President John F. Kennedy in 1963 generated another name change. National sentiment following the president's death prompted the administration to ask the board to rename Cape Canaveral in Florida as Cape Kennedy. Many people believed that the former president's support of U.S. space activities, largely centered at Cape Canaveral, justified an action to honor the former chief executive. The board responded by approving the change. With time, however, opposing views arose. One major concern was that the state of Florida had not been consulted. In addition, there were countless local and national comments disputing the action. Significantly, the name "Cape Canaveral," which identified a lowland on the eastern coast of Florida, was one of the earliest names recorded in the United States and therefore should be preserved. The issue came before the board again and in October 1973 it acted to restore the original name to the feature. The space station in that area, however, was named Kennedy Space Center.

Daugherty or Dougherty in Texas

A dispute about a name can arise over a single letter. In 1984 the board received a letter from a woman in Texas objecting to the board's agreement three years earlier to follow a state move that the name of her town should be Dougherty, not Daugherty. The state decision followed a review of comments from citizens in that town who presented information favoring the name Dougherty. The issue disturbed the woman, whose last name was Daugherty and whose ancestors were important to the town's settlement. Satisfied with the state's decision, however, the board took no further action.[3]

SOME DISPUTES IN FOREIGN AREAS

In 1995 and 1996 the name "Macedonia" was often in the headlines. It was the short form of a new entity, the Republic of Macedonia, which had been the Yugoslav Republic of Macedonia when part of the former Yugoslavia. When the name "Republic of Macedonia" was proposed, Greece raised an immediate protest on the grounds that its history and culture were firmly attached to the name "Macedonia." Its position was that while the name could be assigned to a republic of another country (as was the case when it was a republic of Yugoslavia), it could not be applied to a new and independent nation. Greece issued formal diplomatic objections, ceased normal commerce and other relations with the entity, and even threatened military action if the name were adopted. At present, several countries have recognized the name "Republic of Macedonia," but out of respect for

Greece, the United States calls it the Former Yugoslav Republic of Macedonia.[4] At the seventh U.N. Conference on Geographical Names in 1998, Greece raised objections to reports, including one submitted by the United States, that had any references to the name "Republic of Macedonia."[5]

The Sea of Japan, the East Sea, the Sea of Korea, or East Sea/Sea of Japan

The seventh U.N. Conference in 1998 heard an argument about the name of a water body that has received press attention. The delegates of Japan, North Korea, and South Korea presented papers about the name of the body of water between the west coast of Japan and the east coast of the Korean Peninsula. The two Koreas objected to the name Sea of Japan, which generally is used. Representatives from those countries presented historical evidence that the name was developed only after military conflicts involving Japan early in this century were settled in its favor. They pointed out that names referring to Korea had long historical usage and proposed approval of the name "East Sea" or "Sea of Korea." As a compromise, they would accept a joint name: East Sea/Sea of Japan. The Japanese delegate circulated materials that demonstrated a long and widely accepted usage of the name Sea of Japan. At the twentieth session of the U.N. Group of Experts on Geographical Names in January 2000, the Republic of Korea requested further study of the case, but, as before, with the willingness to apply two names, East Sea/Sea of Japan, as an acceptable conclusion. In this instance, Japan expressed no views. The Democratic People's Republic of Korea was not present. As was the case at the 1992 U.N. Conference on Geographical Names during a discussion about other disputed names, the chair of the session noted the body could not make any decisions about names. Some delegates said perhaps the only body that could offer a view was the International Hydrographic Office (IHO), which has developed standards for hydrographic charting and also has published a book with names of oceans and seas. Such names, however, were selected on the basis of undisputed nomenclature, since the IHO has no authority to settle international names disputes. Since the cited countries, plus a section of Russia, border the sea, however, a name IHO might agree to use could well become acceptable to all countries. Delegates from South Korea attended BGN meetings in Washington D.C. in April 2000 to present their position on a dual name. It is BGN policy that maps can add a second or variant name after the approved name. Some BGN agencies may now be ready to apply the name East Sea/Sea of Japan.

Names in Israel and Countries of the Middle East

Although Jordan and Israel now have a mutual peace accord, a few years ago they argued about names of places common to the history of both nations. At the sixth U.N. Conference on Geographical Names in 1992, Israel protested that Jordan did not honor the ancient Hebrew names of various places in Jordan. Jordan countered by stating the names were of Arabic origin. Furthermore, in a working paper submitted to the conference, Jordan included as a recommendation the need to make "the UN and UNESCO aware of the fact that the Israeli authorities in the occupied territories are trying to efface and abolish the ever existing Arabic names of locations in the occupied territories in Palestine and replace them with Hebrew names."[6] Since U.N. policy did not permit the conference to become involved in names disputes, each side was requested to submit its point of view as a matter of record only.

A related situation occurred in connection with a paper circulated by Israel at the seventh U.N. Conference in 1998. The paper referred to that country's action to prepare exonyms for use by countries having different languages.[7] This was part of a recommendation that all nations should prepare exonyms for external use in order to reduce international confusion about name forms. The paper had a list of twenty-seven names, including Dead Sea, Jerusalem, and Jordan River (in those spellings), along with the romanized Hebrew version and the names as written in Hebrew. Delegates from Iran and Morocco objected strongly to the list, noting that since most if not all names also had Arabic spellings, such spellings should carry equal weight in any international publication. The Israeli delegate agreed that Arabic names should also be printed. Pursuant to this position, the same delegate at the twentieth session of the UNGEGN informed the author that highway signs giving directions and distances to major places in Israel and adjacent countries showed their names in Hebrew, Arabic, and English.[8] Having such names would conform to Israel's requirement to use Hebrew for its place names, would placate Arab countries that believed that many places also were entitled to Arabic names, and would accommodate English-speaking tourists unable to recognize either Hebrew or Arabic.

West Bank or Judea and Samaria

About twelve years ago, a private U.S. citizen requested the board to drop the name "West Bank" in favor of the names "Judea" and "Samaria" for virtually the same area. He noted that the recommended names were of biblical origin and referred to sources favoring the position. Support from Israeli officials was also cited. The board did not study the case, since the matter was within the jurisdiction of the Department of State. That agency took no action. BGN records show, in any case, that as far back as 1967 the name "West Bank" was used by various publications and organizations and had become a standard reference. The BGN Foreign Names Committee approved the name as a region in 1971 and as a "territory" in 1986.

Cyprus

In Cyprus, name disputes relate to the interests of two ethnic groups—the Greek Cypriots and the Turkish Cypriots. Strife between these elements generated action by Turkey in 1974 to occupy roughly the northern third of the island. The result is that the Turkish Cypriots have de facto possession of that segment. The Greek Cypriots administer the only internationally recognized form of a government relevant to the island, the Republic of Cyprus. In 1983, Turkey proclaimed independence for "The Turkish Republic of Northern Cyprus," which is recognized only by Turkey. The impact of these actions on place names has been significant. Both groups claim the other has attempted to change forms of names in their territories with resulting damage to the cultural heritages of the respective populations. The two entities are continuing their efforts to resolve their differences.

NAMES COMMON TO THE UNITED STATES AND CANADA

The United States and Canada are not involved in name disputes having political implications similar to those described above, but the countries are aware that different names are applied to features crossing their common boundary. Meetings of appropriate national bodies have attempted to accept a single name for such features in order to eliminate possible confusion. Factors they considered included the percentage of the length and the size of features or their watersheds on each side, the age of the given names and how they may be parts of names of other nearby features, and the elevations of physical features. The goal was to select a name used by the country in which the major dimensions of a feature were located. Although there was some initial success in resolving the dual naming problem, no names were changed. One difficulty was the inability to identify the defined percentages of areas and lengths to a mutually satisfactory degree. Such dual naming is common to many international features in other parts of the world.

NOTES

1. Information about Mount McKinley and Mount Rainier is based partly on an article by Donald Orth, "The Story of the Naming of Mount Rainier and Other Domestic Names Activities of the U.S. Board on Geographic Names," which appeared in a festschrift for Meredith F. Burrill in *Names*, December 1984. Mr. Orth is former executive secretary of the board's Domestic Names Committee.

2. Curiously, two suburbs along the northeastern boundary of Washington, D.C., are named Takoma Park and Mount Rainier. In 1883, a developer bought about 100 acres along a rail line and gave it the name "Takoma," related to the Indian word "tacoma," said to mean something high but spelled with a *k* instead of a *c* to differentiate it from the settlement of Tacoma in the state of Washington. The developer

later added "Park" to the name. Several years earlier, army engineers from Seattle, Washington, surveyed an area approximately 100 acres in size, presumably predicting a market for settlements, and named it Mount Rainier in honor of the mountain near Seattle. They thought such names would attract property buyers.

3. It is not unusual that the board hears from people who dispute its decisions. With few exceptions, however, the board supports the actions of state-naming authorities.

4. U.S. ties with Greece through NATO and views of various U.S. citizens influenced the decision.

5. At the seventh U.N. Conference on Geographical Names in 1998, a document on country names submitted by a conference committee referred to "Macedonian Cyrillic" in connection with a romanization system applicable to the language of the Former Yugoslav Republic of Macedonia (FYROM). Greece circulated comments that since negotiations on the eventual name of FYROM were still under way, it was improper to use the name "Macedonia" in any context. The cited reference relates, however, not to a U.N. publication but to a U.S. publication that illustrates BGN/PCGN romanization systems and refers to the "Romanization System for Macedonian." Use of the term "Macedonian" does not imply U.S. approval of "Macedonia" as a country name nor does it lessen the validity of the cited system for the language officially used in the FYROM.

6. *Working Paper no.* 1, 20 August 1992, submitted by Jordan to the sixth U.N. Conference on the Standardization of Geographical Names, New York, August 25–September 3, 1992.

7. "Donor-Recommended Exonyms for Historical Topographic Features in Israel," E/Conf.91/L.10, November 7, 1997. Chapter 11, "The Terminology of Names," gives definitions of the term "exonym." The same reference shows that a U.N. glossary of terminology and the cited paper have different definitions for the word.

8. Information provided by the delegate from Israel, Naftali Kadmon. It is a general practice to refer to "English" as the written language for names printed in the Roman alphabet when such names are based to a large extent on naming practices of English-speaking populations. There is evidence that international travel brochures and related publications issued by countries with non-Roman alphabet languages that cover nations other than those speaking English will spell names according to English-language practices. The value of English to tourism is thus evident.

20

Unusual and Unacceptable Names

U nusual, comical, or otherwise interesting or unacceptable place names are quickly noticed. The United States has many such names, as a scan of maps reveals, and they can be grouped according to categories. One kind of name is biblical, which includes dozens of places with names such as Bethlehem, Nazareth, or Zion. Other categories include names ending in "ville" or names that are feminine or masculine. Long and short names fall in other categories. The origins of some names can be traced to the fact that a person may name a feature for a reason other than to provide a formal identification. Otherwise, a name may acquire a humorous or unusual nature with time as people interpret names differently. In other cases, a name may be misspelled and thus acquire a meaning different from the original intent, or subsequent research may reveal an unsuspected context. Growing attention is being given to names that have become unacceptable. There are numerous publications that deal with these kinds of names.[1] This chapter describes a few place names that fall in several categories.

LONG NAMES

For some time, the name of a town in Wales was thought to be the world's longest. It is

Llanfairpwllgwyngllgogerychwyrndrobwllllandysiliogogogoch

which means "St. Mary's Church by the pool of the white hazel trees, near the rapid whirlpool, by the red cave of the church of St. Tysilio." Its short form is "Llanfair." Another long name is a lake in Massachusetts:

Chargoggagoggmanchagogchaubungagungamog

Local Indians gave the lake its name, which is said to mean "you fish on that side, I'll fish on this side, and we'll both fish in the middle." A shorter version is Lake Chaubungagungamog and the conventional name is Lake Webster.

The name of a hill in New Zealand has attained publicity as the longest:

Tuamatawhataktankihangakoauauotamateaturipukakapikimaunga-
   horonukuokaiwhenuakitanatahu

which means, "the place where Tamateakokai-whenua—the man with the big knees who slid, climbed, and swallowed mountains, known as land eater—played his flute to his loved one." One local expert comments that the official name is a bit shorter but another claims the given version is correct.[2] Such names are interesting but may be too long to appear on any but large-scale maps; alternative or conventional names or abbreviations may be used instead. It is most likely they are, however, fully spelled out on local tourist signs or postcards. Long names seem to reflect practices associated with cultures having no written languages. Chapter 1, " What Is a Place Name?" also discusses such naming customs.

SHORT NAMES

A glance at names in an atlas or a gazetteer will find short names. Such names normally are those appearing on specified maps of a relatively large scale, which carry names of small features. Other maps may omit features of that magnitude and thus may omit those names. There are many examples of short names. In Switzerland, the Aa River is famous because its name has only two letters. In the vicinity of Ahaus in northwestern Germany, a small stream is identified by a sign near a bridge. Its apparent name is a single letter: "A."

NAMES WITH COMMON ENDINGS

Cultural factors can affect the terms used as the last elements of names. In the United States, the many names ending with "ville" indicate a common view that this French word for "town" added to a name would give a place a certain level of sophistication. There are 7,865 places in the United States with that element. Pennsylvania has 980, New York has 625, New Jersey has 424, and Indiana, Kentucky, and Virginia each have more than 300.[3] "Boro" is common in the United States because it has a link to the English word for a settled place, "borough," and was seen by namers as useful. In some areas the term "burg" is a common part of names because some people of German background who named communities felt it appropriate to use that term, which in German means a populated place. Elsewhere, "burg" may be an abbreviation of "burgh," a term found in Scotland.[4]

NAMES WITH UNUSUAL ORIGINS

A few names can illustrate the interesting circumstances of their origins. At a time when railroads were the principal mode of travel for any but short distances, trains periodically stopped at stations where their engines could take on coal and water. In California, one railroad named its coaling stations by letters, such as A, B, C, and so on. Eventually, Coaling Station A became the site of a permanent settlement. Its name became "Coalinga." The town still exists.

A place in West Virginia has the name "Pinch." Its original name was "Pinchgut" and was given by lumberers over a century ago whose sparse income brought near starvation conditions that caused them to tighten their belts, thus pinching their guts.

In April 2000, the BGN Domestic Names Committee reported a proposal in Oregon to rename a community to reflect an effort by an Internet company to gain publicity. The current name is Halfway, and the proposed new name is Half.com. No formal action has yet taken place.

A Canadian name with an interesting origin is Whistler, a resort town in British Colombia. A groundhog in the vicinity has the name "whistler" because its call is a high squeal. It seemed appropriate to give that name to the community.

A REQUEST FOR AN ANIMAL-FRIENDLY PLACE NAME

A radio report in the United States in September 1996 said that PETA (People for the Ethical Treatment of Animals) requested the mayor of the town Fishkill in New York to change the town's name to Fishsave. The action would ostensibly promote PETA's position that catching and killing fish is unethical. The mayor informed PETA that the term "kill" originated among early Dutch settlers who used that word to identify rivers or streams. The request was withdrawn.

HOW MANY GREAT LAKES ARE THERE?

Lake Ontario, Lake Erie, Lake Huron, Lake Michigan, and Lake Superior in the United States (also partially bordering on Canada) are collectively called the Great Lakes. It is a regional name with widespread usage. Early in 1998, the U.S. Congress approved a bill to add Lake Champlain in Vermont as a sixth Great Lake. The rationale given by the sponsor of the measure, a senator from that state, was that its geological basis and its ecology were similar to the other Great Lakes. Another important factor was that states with frontage on the Great Lakes or on the open seas qualified for federal support for marine studies. Thus Vermont could also benefit. The reactions were swift. The news media picking up the story indicated that voices in Congress subsequently expressed doubt and that private mapping companies would not comply. Although the U.S. Board on Geographic Names is the national authority for approving names appearing on official maps, Congress does have authority to name standard features without further board action. Following overwhelming public opinion against the recommended name, the case was removed from the agenda.

COMMON NAMES IN THE UNITED STATES

A BGN gazetteer of the United States shows that Lincoln and Washington are the two most common place names in America.[5] Clearly, they are "patronyms," as is the case with many other names, and were given by settlers seeking appropriate names for places. Half or more of such names are for counties, with the remaining applied to county seats, settlements, lakes, and other features. There are literally hundreds of names with adjectival terms such as "big," "long," or "new." For example, more than 600 features have "big" as the initial word of their names. Referring only to one kind of feature, "lake," there are eight places named Big Lake, which include water bodies as well as settlements.

THE SHORTEST RIVER IN UNITED STATES

In 1988, pupils of a grade school in Montana asked the board to approve a name for what they claimed was the shortest river in the country—201 feet—and proposed the name "Roe River." The board accepted the name in the same year. Shortly afterward, a names expert from Oregon noted that a river in his state having the name "D" was only 120 feet. No formal action followed. These small sizes indicate that generic terms have no stipulated dimensions. At a meeting of the board in January 1999, the Domestic Names Committee reported that various states are now having contests in public schools to find and name the shortest river.

WHERE IS A TOWN CALLED LAKE WOBEGON?

The public is well aware of a town named Lake Wobegon in Minnesota, a place that has played an integral part in the popular narrations and publications of Garrison Keillor. His book *Lake Wobegon Days*[6] quotes a report, apparently from an official state publication, that describes the town Lake Wobegon and the adjacent lake (with the same name) that is fed by the Lake Wobegon River. According to the book, the term "wobegon" is evidently tied to the pronunciation of an Indian phrase that meant "here we are" or "we sat all day in the rain waiting for (you)." The text also notes with an aura of historical authenticity that an Italian explorer thought the river was the true headwaters of the Mississippi but soon concluded it was not. Ostensibly quoting local citizens, Keillor adds, "What has made so many others look at us and think: *It doesn't start here?*" His books and radio programs often portray inhabitants of the town and environs as down-to-earth Americans lacking a degree of sophistication. The name itself seems to be connected with another word: "woebegone."

According to authorities in Minnesota, however, there is no town or lake named Lake Wobegon and never was.[7] Action in the 1980s to apply that name, which had become famous, to an unnamed lake in Minnesota was rescinded because officials may have thought the name was copyrighted. The subject feature then received the name Lake Cavanaugh. There is, however, a Wabegon Lake whose name BGN accepted in 1941 after earlier approval by Minnesota state authorities. Apparently Keillor, being familiar with Minnesota, knew of that name and adopted a version of it for his fictional although realistic stories about life in a community of his imagination. In 1996, the board approved the name, Wahbegon Lake, in response to action by the state to apply that name to replace Squaw Lake, which it decreed was not suitable.[8]

A PUNCTUATION MARK MEANS A NEW NAME?

In 1986, the city manager of Hamilton, Ohio, asked the board to accept a town council decision to approve its name with an added exclamation point: Hamilton! Local authorities felt the modified name would upgrade its image and be attractive to residents and businesses. The board responded it would not consider the request because punctuation marks cannot be part of place names. The board noted it could not dictate against the use of Hamilton! for local correspondence or promotion. Consequently, the official name remains as before—without an exclamation point.

THE LAST NAME IN ANY GAZETTEER

The board acted in 1984 to accept a decision of a California advisory committee to name a resort Zzyzx. The reason for the name was to give it the distinction of always being last in any list of names in alphabetical order.

UNACCEPTABLE NAMES IN THE UNITED STATES

Many groups worldwide have sought to change place names they see as unacceptable for a variety of reasons. Native populations want to restore their original names if they have been replaced by newer nomenclature. Other minority groups want to eliminate names they find racially offensive. Desired changes of names can also reflect other factors. Cases in the United States described below also might be "disputed" names, but the circumstances are different.

A growing tide of complaints about names considered unacceptable has arisen. In virtually all instances, the problem has been with the "specific term" of a name, that is, the element that identifies a physical feature (referred to as a "generic term"). As an example, BGN many years ago did not approve a place name in Oregon because its specific term was considered obscene. The name was Whorehouse Meadow, given by early settlers who believed that some members of the native population had conducted certain activities at a particular place. The state reacted to the board's policy and changed the name to Naughty Girl Meadow, a decision the board approved. In the 1960s, a new outlook by the population of Oregon moved the state to approve the old name. The board reviewed its decision and accepted the change.

More recently, broad categories of names have become unacceptable. For a number of years, any place name with the term "nigger" has not been printed on any official U.S. map because the word was seen as totally pejorative and unacceptable and therefore not suitable for any U.S. product. The words "black" and "negro" were substituted, but in time those terms also were seen as objectionable and efforts to ban them arose with various degrees of effectiveness. The origins and meanings of some such terms, however, need to be understood, an analytical process that can modify popular concepts. For example, the term "black" could have referred to a person with that name, in which case there was no demonstrable tie to a racial origin. Or the word could have been selected to identify the dark color of a river, lake, or another kind of feature. The word "negro" can also have a nonpejorative meaning. In the Spanish language, it means "black" as in the color. With an increasing tendency to honor Spanish names in the United States, the word "negro" associated with a name could be acceptable where shown to have such an origin. The U.S. gazetteer cited in note 5 of this chapter lists some seventy-five features whose names include "black." There are fourteen Black Rivers. Thus the board and its counterparts in the states are facing challenges regarding the treatment of these and similar kinds of names. A related issue is that other kinds of names could be deemed objectionable if they honor a person who was a slaveholder. In the state of Louisiana, a public school named for George Washington changed that name in 1999 because he possessed slaves. Applying this policy to place names may be a logical next step.

In the United States, "squaw" has become another unacceptable term. Although long understood to mean a female Indian, many people object to it because they say it is derived from a word French traders used to describe the private parts of such women. Now there is action to change place names that have that word, but the process may be slow because there are more than 1,000 such place names, many of them being small features. Examples of major features are Squaw Lake in New Hampshire, Squaw Creek in Iowa, and Squaw Valley in California. Minnesota has taken a major step by ruling that the term "squaw" in any name in that state be replaced. The state has in fact declared the word unsuitable for use in any official state publication. One county, responding to what it believed was an improper state policy, offered to change the name of a Squaw Lake to "Politically Incorrect Lake." State authorities did not accept the recommended substitute. The Domestic Names Committee reported in April 2000 that the state of Maine was considering banning the word "squaw" from place names.

Similar actions related to attempts to recognize Indian place names in addition to, or in place of, those imposed by settlers or others have already been noted. Hispanic names are currently receiving greater attention in the United States. Since many such names in the records may not be officially recognized, efforts to restore them could produce disputes with names later applied.

If demands for giving greater attention to "native" names are common in the future, it seems logical that nomenclature bestowed by people from Europe may be increasingly challenged. In addition, given the U.S. requirement to enforce the concept of separation of church and state, names with any religious connotation may be subject to review. Thus names as Bethlehem, Corpus Christi, Santa Fe, and St. Louis may be declared improper. The list of such names is sizable. In the United States there are about 280 place names with "St." or "Saint" and some 370 with "San" or "Santa." A policy banning such names could counter another possible program to restore Hispanic names or to give them a status equal to that of names later created. Despite the new standing of Hispanic names, any with the terms "San" or "Santa" thus might prove to be unacceptable.

The board also listens to comments from its foreign associates about unacceptable names. Geisha Guyots was listed in records of the board's Advisory Committee on Undersea Features for a number of years. (A guyot is a mountainlike undersea feature.) In the past decade, Japanese colleagues of that committee registered their view that the name was improper because it actually referred to a woman whose character was far worse than that of a dancing girl the term traditionally indicated. The committee's 1992 gazetteer shows that the currently approved name is "Japanese Guyots."

NOTES

1. Books dealing with unusual place names include Leslie Dunkling, *The Guinness Book of Names* (Enfield, U.K.: Guinness, 1989); David Jouris, *All Over the Map: An Extraordinary Atlas of the United States* (Berkeley, Calif.: Ten Speed, 1994); Derek Nelson, *Off the Map: The Curious Histories of Place-Names* (New York: Kodansha International, 1997).

2. Maurice D. Bartlett, personal correspondence with author, April 16, 1999. Bartlett retired after a career in cartography and land analysis in New Zealand. He provides a map and a postcard with the full name. He also states that the cited name may be longer than the correct version. Wendy Shaw, secretary of New Zealand Geographic Board, personal correspondence with author, October 11, 1998, confirms that the name as spelled is correct and by official rules must appear that way on maps or relevant documents.

3. Information from a printout of digital files compiled by the Board on Geographic Names Domestic Names Committee in 1992.

4. BGN ruled early in the twentieth century that place names should be simplified wherever possible. One result was that the term "borough" after any name should be shortened to "boro." Many if not most places with such names made the change. With time, however, some restored the original spelling. The United States now has fourteen places called Marlborough and seventeen called Marlboro. Massachusetts has two places, one with each name. Similarly, the term "burgh" should be shortened to "burg." There was substantial compliance, but the city of Pittsburgh in Pennsylvania retained its original spelling.

5. *The National Gazetteer of the United States of America*, concise ed., U.S. Geological Survey Professional Paper 1200-US (Washington, D.C.: U.S. Government Printing Office, 1990). This publication contains 42,000 place names of major features in the United States and its territories. It has all names in alphabetical order, identifies feature types according to designations, shows the county and state for each place, gives latitude and longitude to the nearest second, and indicates the highest elevation for features having a finite location.

6. Garrison Keillor, *Lake Wobegon Days* (New York: Viking Penquin, 1985), 1–2.

7. Glen Yakel, supervisor, Hydrographics Unit, Minnesota Department of Natural Resources. Personal correspondence, March 28, 1997. (In an article in the December 2000 issue of *National Geographic*, "In Search of Lake Wobegon," Garrison Keillor states the name is fictional.)

8. It would appear unusual for state authorities to approve two names whose spellings are nearly identical, particularly given the existence of a third name that, although fictional, is virtually the same.

Afterword

Any book dealing with place names on a global or even a local basis is subject to change. Since this work was completed, a number of nations have created new administrative territories. For example, Venezuela formed an administrative territory called Vargas along its Caribbean coast. India has three new states, bringing the total number to twenty-five. Also, Sri Lanka has formed an additional unit to share its governing tasks with two other places. Any national or international program seeking to keep abreast of such developments requires careful research of official materials disseminated by countries. From the U.S. point of view, such an effort has long been the task of its Board on Geographic Names, and meetings of its committees regularly include this category of details.

As already noted, many nations are dealing with language situations. The evidence that previously disregarded languages are to be given validity and thus recognized as part of national linguistic patterns is resulting in what some see as factors complicating common communications. Yet the phenomenon is realistic. Nevertheless, there are indications that nations are resisting what otherwise might be seen as culturally inevitable developments. France resists the use of English terminology in areas such as the Internet. Germany is looking at ways to enhance not only its traditional grammar and vocabulary but also the formerly standard fraktur typeface. These dynamics can affect geographic names. All nations will have to deal with such developments if they are to implement the universally recognized requirement to have accurate and consistent names of places. Fulfilling that requirement is, of course, necessary to ensure effective communications.

Yet in spite of the broad spectrum of changing place names—foreign and domestic—that reflect a range of factors, there remains substantial evidence of how they have become, and are likely to remain, significant elements of our everyday life. Logically, the ongoing flood of new place names will require continuous revision of maps, atlases, gazetteers, and related reference works. Logic may also dictate that this book also be periodically revised. In any case, the author hopes that this work will provide its readers with a useful picture of how place names define the world—and more.

Appendix A

Common Words Derived from Place Names

Angora—A kind of wool from the Angora goat, named for its place of origin in Ankara, Turkey

Aran—A sweater originating on an island with that name off the west coast of Ireland

Argyle—A sweater or sock design that carries the name of a province in Scotland, where the design originated

Baloney—A kind of ground meat in sausage form probably originating in Bologna, Italy

Bayonet—A short sword designed to be attached to a rifle, first made in Bayonne, France

Bordeaux—Red and white wines named for a seaport in France

Bourbon—A distilled liquor originating in the county of Bourbon in Kentucky

Brie—A cheese made in an area with that name in France

Brougham—A style of automobile named after Lord Brougham, ca. 1850, whose title reflected an area in England with that name

Brussels sprouts—Named for the Belgian town of Brussels

Burgundy—A red, somewhat heavy wine from the French province of Burgundy

Calvados—A liquor processed in a department with that name in France

Cambric—A fine cotton fabric originating in the French town of Cambrai

Camembert—A cheese made near a town in Normandy, France, with that name

Canteloupe—A fruit, although originating in India, that got its English name from the Italian town Cantalupo, where it was first successfully grown in Europe

Cardigan—A kind of sweater or jacket named after the Earl of Cardigan, whose title referred to an area in England

Cashmere—A fine wool from goats raised in Kashmir

Champagne—An effervescent wine originally produced in Champagne, a region in northeastern France

Cherry—A fruit named after the city of Cesarus, once part of ancient Greece

Cheviot—A kind of wool spun from sheep in the Cheviot Hills in England

Chianti—A wine produced at or near the Chianti Mountains in Italy

China—Porcelain originally developed in China, now a common synonym for porcelain, such as plates, saucers, and other tableware

Coach—A horse-drawn vehicle the name of which was also used in relation to automobiles; the name based on place of origin, Kocs, Hungary

Cologne—A scented liquid for body or face use originating in Cologne, the French name for Köln, Germany. The formal name is eau de cologne. (The German name is Kölnisches Wasser.)

Copper—Derived from the name "Cyprus," where the metal may originally have been found

Cordovan—A kind of leather originating in Cordoba, Spain, mainly used for shoes

Cravat—Related to scarves worn by soldiers from Croatia serving as French mercenaries in the 1700s. The French word for a person from Croatia is Cravate, the name eventually associated with the scarves. The article also became popular in France and later became a common international piece of attire sometimes still called a cravat (in English), but normally called a "necktie" or "tie."

Currants—Named after Corinth in Greece

Damask—A fine fabric originating in or near a Syrian city, Damascus

Denim—A coarse, colored cotton fabric first made in Nîmes, a French town; being from that city, it became known as "de Nîmes," eventually becoming denim.

Derby—A hat popularized by the English Lord Derby

Dollar—A coin of silver from a mine near a place in Germany called Joachimstaler, which, if translated, would mean Joachim Valley. The coin eventually was called "Taler," based on the German word for valley. Later the term was applied to the U.S. coin known as a "dollar."

Edam—A cheese associated with a place by that name in Holland famous for its cheese market

Fez—A kind of brimless hat originating in Fez, Morocco

Frankfurter—A kind of sausage named for Frankfurt, Germany

Gauze—A thin cloth first made in Gaza

Gouda—A cheese made at a place with that name in Holland

Gruyere—A cheese made at a place with that name in Switzerland

Gypsy—A person from an ethnic group migrating from India but thought to be of Egyptian origin and hence given a name associated with that country

Hamburg or hamburger—Named for Hamburg, Germany, its probable place of origin

Homburg—A man's hat first made in Bad Homburg, a German resort place

Java—A kind of coffee from Indonesia; applied in a popular sense to any coffee

Jeans—As in "blue jeans," a coarse fabric originally produced in, and acquiring its name from, Genoa, Italy

Leyden jar—First manufactured in Leyden, Holland

Limousine—A kind of automobile named for a style of hooded coat worn in the French province of Limousin

Mackinaw—A heavy jacket originating on Mackinac Island in Michigan. The different spelling is probably due to a pronunciation variation.

Madras—A cloth that owes its name to the port of the same name in India. (The place name Madras is now Chennin.)

Manila—A kind of heavy paper used for envelopes named for Manila, capital of the Philippines

Meander—To take a path that twists and turns, from the Meander River in Turkey, whose course and currents are irregular

Millinery—Reference to a fine fabric manufactured in Milan, Italy, and imported by England, where the name evolved from Milaner to millinery

Mocha—A kind of coffee named for a port in the Republic of Yemen but also used as a popular name for any coffee

Muslin—A cotton cloth originating in Mosul, Iraq

Panama hat—A fabric hat originally produced in Ecuador but given the name Panama, probably for sales appeal

Parmesan—A cheese from the Italian town of Parma

Parmigiana—Coated with parmesan cheese and cooked; also from Parma

Pheasant—A bird taking its name from the Phases River, which flowed into the Black Sea

Pilsner—A kind of beer originally brewed in Plzeň (former German name: Pilsen) in the Czech Republic

Port—A kind of wine from a port town, O Porto, in Portugal

Rhinestone—A semiprecious stone named for its original source area, the Rhine River

Roquefort—A cheese named after a town in France with the same name

Samaritan—A good person, from the biblical story of the good citizen of Samaria, an ancient province of Israel

Sardine—A kind of fish named after Sardinia, an Italian island administrative division in the Mediterranean

Sardonic—Named after a certain plant found in Sardinia

Sauterne—A wine from a district with that name near Bordeaux in France

Scotch—A liquor originating in Scotland

Sedan—A type of automobile originating in the French city with the same name

Sherry—A kind of wine taking its name from Jerez de la Frontera, a seaport in Cadiz, a Spanish province

Shetland pony—Named for its original breeding ground, the Shetland Islands off the coast of Scotland

Shropshire—A breed of sheep raised in a region with that name between England and Wales

Tabasco—A spicy seasoning liquid taking its name from a river and a state in Mexico

Tangerine—An orangelike fruit first grown in Asia but possibly imported via the Moroccan port of Tangier for further distribution

Turkey—The name given to a bird by early American settlers who thought it resembled a European fighting bird known to be from Turkey

Turquoise—A kind of semiprecious stone with a bluish color based on the name of the country Turkey, where the gems were first found

Tuxedo—A formal suit based on the name Tuxedo Park in New York, where wealthy landowners specialized in a new style of dinner coats that became popular for formal occasions

Tweed—A coarse woolen fabric made near the River Tweed in Scotland

Wiener—A sausage kind of meat dish originating in Vienna (the German name is Wien), Austria, or at least given that name for purposes of association

Appendix B

Independent States of the World

Short-form name[1]	Long-form name	Code[2]	Capital[3]
Afghanistan	Islamic State of Afghanistan	AF	Kabul
Albania	Republic of Albania	AL	Tirana
Algeria	Democratic and Popular Republic of Algeria	AG	Algiers
Andorra	Principality of Andorra	AN	Andorra la Vella
Angola*	Republic of Angola	AO	Luanda
Antigua and Barbuda	(none)	AC	Saint John's
Argentina	Argentine Republic	AR	Buenos Aires
Armenia	Republic of Armenia	AM	Yerevan
Australia	Commonwealth of Australia	AS	Canberra
Austria	Republic of Austria	AU	Vienna
Azerbaijan	Azerbaijani Republic	AJ	Baku
Bahamas, The	Commonwealth of The Bahamas	BF	Nassau
Bahrain	State of Bahrain	BA	Manama
Bangladesh	People's Republic of Bangladesh	BG	Dhaka
Barbados	(none)	BB	Bridgetown
Belarus	Republic of Belarus	BO	Minsk
Belgium	Kingdom of Belgium	BE	Brussels

159

Belize	(none)	BH	Belmopan
Benin	Republic of Benin	BN	Porto-Novo
Bhutan	Republic of Bhutan	BT	Thimpu
Bolivia	Republic of Bolivia	BL	La Paz (administrative)
			Sucre (legislative/judiciary)
Bosnia and Herzegovina	(none)	BK	Sarajevo
Botswana	Republic of Botswana	BC	Gaborone
Brazil	Federative Republic of Brazil	BR	Brasilia
Brunei	Negara Brunei Darussalam	BX	Bandar Seri Begawan
Bulgaria	Republic of Bulgaria	BU	Sofia
Burkina Faso	Burkina Faso	UV	Ouagadougou
Burma	Union of Burma	BM	Rangoon
Burundi	Republic of Burundi	BY	Bujumbura
Cambodia	Kingdom of Cambodia	CB	Phnom Penh
Cameroon	Republic of Cameroon	CM	Yaoundé
Canada	(none)	CA	Ottawa
Cape Verde	Republic of Cape Verde	CV	Praia
Central African Republic	Central African Republic	CT	Bangui
Chad	Republic of Chad	CD	N'Djamena
Chile	Republic of Chile	CI	Santiago
China	People's Republic of China	CH	Beijing
Colombia	Republic of Colombia	CO	Bogotá
Comoros	Federal Islamic Republic of the Comoros	CN	Moroni
Congo (Brazzaville)	Republic of the Congo	CF	Brazzaville
Congo (Kinshasa)	Democratic Republic of the Congo	CG	Kinshasa
Costa Rica	Republic of Costa Rica	CS	San José
Côte d'Ivoire (Ivory Coast)	Republic of Côte d'Ivoire	IV	Yamoussoukro
Croatia	Republic of Croatia	HR	Zagreb
Cuba+	Republic of Cuba	CU	Havana
Cyprus	Republic of Cyprus	CY	Nicosia
Czech Republic	Czech Republic	EZ	Prague
Denmark	Kingdom of Denmark	DA	Copenhagen
Djibouti	Republic of Djibouti	DJ	Djibouti
Dominica	Commonwealth of Dominica	DO	Roseau

Dominican Republic	Dominican Republic	DR	Santo Domingo
Ecuador	Republic of Ecuador	EC	Quito
Egypt	Arab Republic of Egypt	EG	Cairo
El Salvador	Republic of El Salvador	ES	San Salvador
Equatorial Guinea	Republic of Equatorial Guinea	EK	Malabo
Eritrea	State of Eritrea	ER	Asmara
Estonia	Republic of Estonia	EN	Tallinn
Ethiopia	Federal Democratic Republic of Ethiopia	ET	Addis Ababa
Fiji	Republic of the Fiji Islands	FJ	Suva
Finland	Republic of Finland	FI	Helsinki
France	French Republic	FR	Paris
Gabon	Gabonese Republic	GB	Libreville
Gambia, The	Republic of The Gambia	GA	Banjul
Georgia	(none)	GG	T'bilisi
Germany	Federal Republic of Germany	GM	Berlin
Ghana	Republic of Ghana	GH	Accra
Greece	Hellenic Republic	GR	Athens
Grenada	(none)	GJ	Saint George's
Guatemala	Republic of Guatemala	GT	Guatemala City
Guinea	Republic of Guinea	GV	Conakry
Guinea-Bissau	Republic of Guinea-Bissau	PU	Bissau
Guyana	Co-operative Republic of Guyana	GY	Georgetown
Haiti	Republic of Haiti	HA	Port-au-Prince
Holy See*	Holy See	VT	Vatican City
Honduras	Republic of Honduras	HO	Tegucigalpa
Hungary	Republic of Hungary	HU	Budapest
Iceland	Republic of Iceland	IC	Reykjavík
India	Republic of India	IN	New Delhi
Indonesia	Republic of Indonesia	ID	Jakarta
Iran+	Islamic Republic of Iran	IR	Tehran
Iraq+	Republic of Iraq	IZ	Baghdad
Ireland	(none)	EI	Dublin
Israel	State of Israel	IS	(In 1950, the Israeli parliament proclaimed Jerusalem as the capital. The United States,

			like some other states, maintains its embassy in Tel Aviv.)
Italy	Italian Republic	IT	Rome
Jamaica	(none)	JM	Kingston
Japan	(none)	JA	Tokyo
Jordan	Hashemite Kingdom of Jordan	JO	Amman
Kazakhstan	Republic of Kazakhstan	KZ	Astana
Kenya	Republic of Kenya	KE	Nairobi
Kiribati*	Republic of Kiribati	KR	Tarawa
Korea, North	Democratic People's Republic of Korea	KN	P'yongyang
Korea, South	Republic of Korea	KS	Seoul
Kuwait	State of Kuwait	KU	Kuwait
Kyrgyzstan	Kyrgyz Republic	KG	Bishkek
Laos	Lao People's Democratic Republic	LA	Vientiane
Latvia	Republic of Latvia	LG	Riga
Lebanon	Lebanese Republic	LE	Beirut
Lesotho	Kingdom of Lesotho	LT	Maseru
Liberia	Republic of Liberia	LI	Monrovia
Libya	Socialist People's Libyan Arab Jamahiriya	LY	Tripoli
Liechtenstein	Principality of Lichtenstein	LS	Vaduz
Lithuania	Republic of Lithuania	LH	Vilnius
Luxembourg	Grand Duchy of Luxembourg	LU	Luxembourg
Macedonia, The Former Yugoslav Republic of	The Former Yugoslav Republic of Macedonia	MK	Skopje
Madagascar	Republic of Madagascar	MA	Antananarivo
Malawi	Republic of Malawi	MI	Lilongwe
Malaysia	(none)	MY	Kuala Lumpur
Maldives	Republic of Maldives	MV	Male
Mali	Republic of Mali	ML	Bamako
Malta	(none)	MT	Valletta
Marshall Islands	Republic of the Marshall Islands	RM	Majuro
Mauritania	Islamic Republic of Mauritania	MR	Nouakchott

Mauritius	Republic of Mauritius	MP	Port Louis
Mexico	United Mexican States	MX	Mexico City
Micronesia, Federated States of	Federated States of Micronesia	FM	Palikir
Moldova	Republic of Moldova	MD	Chisinau
Monaco	Principality of Monaco	MN	Monaco
Mongolia	(none)	MG	Ulaanbaatar
Morocco	Kingdom of Morocco	MO	Rabat
Mozambique	Republic of Mozambique	MZ	Maputo
Namibia	Republic of Namibia	WA	Windhoek
Nauru*	Republic of Nauru	NR	Yaren District (no capital city)
Nepal	Kingdom of Nepal	NP	Kathmandu
Netherlands	Kingdom of the Netherlands	NL	Amsterdam The Hague (seat of government)
New Zealand	(none)	NZ	Wellington
Nicaragua	Republic of Nicaragua	NU	Managua
Niger	Republic of Niger	NG	Niamey
Nigeria	Federal Republic of Nigeria	NI	Abuja
Norway	Kingdom of Norway	NO	Oslo
Oman	Sultanate of Oman	MU	Muscat
Pakistan	Islamic Republic of Pakistan	PK	Islamabad
Palau	Republic of Palau	PS	Koror
Panama	Republic of Panama	PM	Panama
Papua New Guinea	Independent State of Papua New Guinea	PP	Port Moresby
Paraguay	Republic of Paraguay	PA	Asunción
Peru	Republic of Peru	PE	Lima
Philippines	Republic of the Philippines	RP	Manila
Poland	Republic of Poland	PL	Warsaw
Portugal	Portuguese Republic	PO	Lisbon
Qatar	State of Qatar	QA	Doha
Romania	(none)	RO	Bucharest
Russia	Russian Federation	RS	Moscow
Rwanda	Rwandese Republic	RW	Kigali
Saint Kitts and Nevis	Federation of Saint Kitts and Nevis	SC	Basseterre
Saint Lucia	(none)	ST	Castries
Saint Vincent and the Grenadines	(none)	VC	Kingstown

Samoa	Independent State of Samoa	WS	Apia
San Marino	Republic of San Marino	SM	San Marino
São Tomé and Principe	Democratic Republic of São Tomé and Príncipe	TP	São Tomé
Saudi Arabia	Kingdom of Saudi Arabia	SA	Riyadh
Senegal	Republic of Senegal	SG	Dakar
Seychelles	Republic of Seychelles	SE	Victoria
Sierra Leone	Republic of Sierra Leone	SL	Freetown
Singapore	Republic of Singapore	SN	Singapore
Slovakia	Slovak Republic	LO	Bratislava
Slovenia	Republic of Slovenia	SI	Ljubljana
Solomon Islands	(none)	BP	Honiara
Somalia	(none)	SO	Mogadishu
South Africa	Republic of South Africa	SF	Pretoria (administrative) Cape Town (legislative) Bloemfontein (judiciary)
Spain	Kingdom of Spain	SP	Madrid
Sri Lanka	Democratic Socialist Republic of Sri Lanka	CE	Colombo
Sudan	Republic of the Sudan	SU	Khartoum
Suriname	Republic of Suriname	NS	Paramaribo
Swaziland	Kingdom of Swaziland	WZ	Mbabane (administrative) Lobamba (legislative)
Sweden	Kingdom of Sweden	SW	Stockholm
Switzerland*	Swiss Confederation	SZ	Bern
Syria	Syrian Arab Republic	SY	Damascus
Tajikistan	Republic of Tajikistan	TI	Dushanbe
Tanzania	United Republic of Tanzania	TZ	Dar es Salaam Dodoma (legislative)
Thailand	Kingdom of Thailand	TH	Bangkok
Togo	Togolese Republic	TO	Lomé
Tonga	Kingdom of Tonga	TN	Nuku'alofa
Trinidad and Tobago	Republic of Trinidad and Tobago	TD	Port-of-Spain
Tunisia	Republic of Tunisia	TS	Tunis
Turkey	Republic of Turkey	TU	Ankara
Turkmenistan	(none)	TX	Ashgabat
Tuvalu*	(none)	TV	Funafuti

Uganda	Republic of Uganda	UG	Kampala
Ukraine	(none)	UP	Kiev
United Arab Emirates	United Arab Emirates	TC	Abu Dhabi
United Kingdom	United Kingdom of Great Britain and Northern Ireland	UK	London
United States	United States of America	US	Washington, D.C.
Uruguay	Oriental Republic of Uruguay	UY	Montevideo
Uzbekistan	Republic of Uzbekistan	UZ	Tashkent
Vanuatu	Republic of Vanuatu	NH	Port-Vila
Venezuela	Bolivarian Republic of Venezuela	VE	Caracas
Vietnam	Socialist Republic of Vietnam	VM	Hanoi
Yemen	Republic of Yemen	YM	Sanaa
Zambia	Republic of Zambia	ZA	Lusaka
Zimbabwe	Republic of Zimbabwe	ZI	Harare

Other territories

Serbia and Montenegro (Although these states and other nations refer to these entities as Yugoslavia, the U.S. continue to regard them as separate entities.)   SR Belgrade (Serbia) MW Podgorica (Montenegro)

Taiwan (With the establishment of diplomatic relations with China in 1979, the U.S. government recognized the People's Republic of China as the sole legal government of China and acknowledged that there is only one China and that Taiwan is part of China. This does not preclude U.S. relations with Taiwan for a variety of purposes. While Taiwan is claimed by both the government of the People's Republic of China and the authorities of Taiwan, the entity is administered by authorities of Taiwan.)   TW Taipei

NOTES

The information displayed in this appendix comes from the U.S. Department of State, January 21, 2000.

1. All nations have diplomatic relations with the United States and are members of the United Nations unless otherwise indicated. Nations marked with * have relations with the United States but are not U.N. members. Nations marked with + have no relations with the United States but are U.N. members.

2. Two-letter codes are elements of the U.S. Federal Information Processing Standards, a system that provides abbreviated yet accurate identification of countries for use in official U.S. communications.

3. Names of capitals either are local names, if in Roman-alphabet countries, conventional names as approved by the U.S. Board on Geographic Names, or, if in non-Roman alphabet countries, are local names spelled in accordance with romanization systems approved by BGN. Such spellings may include diacritics and special markings relevant to such romanized names.

Appendix C

Structure and Working Procedures of the U.S. Board on Geographic Names

STRUCTURE

Agencies Responsible for Board Membership

Department of Agriculture
Department of the Interior
Department of Commerce
U.S. Postal Service
U.S. Government Printing Office

Library of Congress
Central Intelligence Agency
Department of Defense
Department of State

Committees

Domestic Names Committee
Department of Agriculture
Department of Commerce
Department of the Interior
Library of Congress

Foreign Names Committee
Central Intelligence Agency
Department of Defense
Department of State

U.S. Government Printing Office
U.S. Postal Service

Advisory Committees

Advisory Committee on Antarctic Names
Advisory Committee on Undersea Features
Advisory Committee on Extraterrestrial Features (being reestablished)

WORKING PROCEDURES

The Board

The board elects a chair and a vice chair generally for two-year terms. Their responsibilities include leading quarterly board meetings, preparing agendas of sessions, preparing minutes for circulation to members, and otherwise representing the BGN at national or international meetings. Working closely with these officers and its committees is the board's executive secretary who serves to coordinate the work of the entire board. He or she is an employee of either the National Imagery and Mapping Agency (NIMA) or the U.S. Geological Survey (USGS), is selected by board members, and is appointed, with concurrence of NIMA and USGS, for an indefinite period of time by the secretary of the interior. The executive secretary also is the executive secretary either of the board's Domestic Names Committee or its Foreign Names Committee. As such, the person has numerous dual responsibilities. Furthermore, the individuals also serve as principal representatives of the board and the United States at meetings of relevant national and international organizations. Quarterly board meetings can include as many as 10 members or deputy members, one or both executive secretaries, staff from the USGS and NIMA, and guests representing other federal agencies and such organizations as the National Geographic Society.

The Domestic Names Committee

The Domestic Names Committee has one member and one or more deputy members from each agency involved with names in the United States and its territories and dependencies. Members elect a chair and a vice chair to run meetings. One major officer is the DNC executive secretary, who is an employee of the USGS and directs the work of the supporting DNC programs. The officer sets agendas, prepares background documentation, provides technical assistance during meetings, reports on committee actions at board meetings, and generally represents the committee in contacts with other organizations, including the press. The committee meets monthly to review names brought before it by its staff or other related matters. The USGS supports the eight-member DNC staff and its production programs. The staff consists of individuals with backgrounds in geography, cartography, and other fields who, as a result of their work on place names, are defined as toponymists.

The committee has a series of principles, policies, and procedures that form the basis of its actions. Most DNC action is to approve proposals for new or recently changed feature names. Such proposals may come from individuals, from state agencies responsible for place names, or from various U.S. agencies and are submitted to the committee as dockets. As an example of DNC procedures, one point defines how to treat a proposal to name a feature in honor of a person. Two conditions usually are required. First, the person must have had a demonstrable relationship with the feature, that is, lived on or near the place for a reasonable period of time. Second, the person must have been dead for at least five years. As circumstances warrant, however, the guidelines are reviewed and possibly modified. One newly accepted regulation was to approve diacritic marks on any names in the United States where linguistically valid. This issue had been on the agenda for many years, but the pro-Anglophone customs precluded diacritics. A few years ago, however, Hawaii requested the right to have its names written as spelled in the local indigenous form and with diacritics where such were normal. The committee sanctioned such names as official and is now examining wider application of other languages in the United States where circumstances warrant.

Another issue arose recently when the governor of Puerto Rico sent the board a letter identifying the principal island as "Island of Puerto Rico and Isla de Puerto Rico." The committee replied that the board's policy on univosity required only a single name, and thus only the name Island of Puerto Rico was acceptable. If Puerto Rico becomes a state, this and other practices regarding "foreign" spellings will be under review.

Reports of committee actions are included in agendas prepared for quarterly board sessions. Normally, the board members do not become involved in discussions about DNC actions but tend to accept them as presented. Committee decisions are also submitted to the secretary of the interior as items of information. Certain issues may require the secretary's approval.

A major program of the Domestic Names Committee is to produce state gazetteers. The work is carried out by the states under contractual arrangement with, and according to specifications of, the committee. As of October 1998, eight state gazetteers had been printed. The advent of digital production, however, has resulted in digital gazetteers covering all states.

As of early 1998, forty-two states had created agencies with responsibilities for naming features or correcting them as required. They basically follow regulations developed by the Domestic Names Committee and usually coordinate their decisions with the national body. However, committee decisions made in behalf of the board apply only to federal products. Names approved at the state level that do not comply with national rules cannot be placed on U.S. maps or charts. Collaborative efforts by state names agencies and the DNC are increasing. Each year, a Western States Names Conference has been attended by representatives of the board, and much useful progress resulted. In 1998, the organization expanded to include all interested states (even though all do not have names committees) and is called the Council of Geographic Names Authorities.

Samples of Actions on Proposals for New Names Submitted to the Domestic Names Committee

Proposal for a New Name—Mine Hole Brook, New York—Acted upon by the Committee on 11 December 1997

Background. This proposal is to make official the name, Mine Hole Brook, for an unnamed 1.8 mile-long tributary of Rondout Creek in the Town of Wawarsing in Ulster County. The name has been in use since the late 19th century and was mentioned in an 1887 county land deed. It also appeared on a 1918 Town of Wawarsing election map and on a 1951 property map, and has been mentioned in two local histories of the area, one of which was recently published by the proponent. The name of the stream was derived from its proximity to an old abandoned mine located near the source of the stream. Four long-time residents of the area interviewed by the proponent verified the name, as did the Ulster County historian. The Town Board of Wawarsing passed a resolution in favor of making this name official, and the New York Committee on Geographic Names recommends approval of this name.

The committee minutes recorded this action:

This proposal is to make official a name which has been in use since the late 19th century for a tributary of Rondout Creek. A motion was made and seconded to approve the name.

Vote: 4 in favor
0 against
0 abstentions

Proposal to Accept a New Name—Oaksmith Island, Alaska—Acted upon by the Committee on 10 July 1997

Background: This proposal was submitted by the grandson of Stanley and Martine Oaksmith to name a small (1.4 acres) island in Tongass Narrows in the

Ketchikan Gateway Borough that his family purchased in 1995. The Oaksmiths arrived in Ketchikan in 1905, and both Mr. and Mrs. Oaksmith were pioneers and business owners in the area. Mr. Oaksmith died in 1942, while Mrs. Oaksmith ran a dress shop until one year before her death at the age of 104 in 1987. The U.S. Forest Service has no objection to the proposal, which is located outside of the Tongass National Forest, but which would appear on USFS Visitor Maps. The Ketchikan Gateway Borough Assembly approved the name for use on local maps at the request of the Oaksmith family. The Alaska Historical Commission (AHC-State Board) denied the proposal since the proponent's grandparents did not have a direct association with the feature, and the AHC did not believe that they had attained regional notoriety. The AHC instructed its staff to advise the proponent that if the name does become part of the local vernacular then they may resubmit the proposal in 7 to 10 years.

The committee minutes record this action:

The proposal to name a small island in Tongass Narrows in the Ketchikan Gateway Borough was submitted by the grandson of Stanley and Martine Oaksmith, who had been pioneers and business owners in the area. A motion was made and seconded not to approve the name.

Vote: 4 in favor
1 opposed
0 abstentions

The proposal was rejected in agreement with the view of the Alaska Historical Commission that the persons being honored did not have a direct association with the feature and had not attained regional notoriety. The vote opposing the rejection was cast by a member who felt the criteria of the committee's Commemorative Naming Policy had been met.

Foreign Names Committee

This committee has one member and one or more deputy members from federal agencies that have direct interests in collecting and processing names of foreign areas. Names processed by the committee are presented by its staff at NIMA, which consists of a group of some thirty-five experts in linguistics, cartography, languages, and geography. Some members were born in other countries and consequently have personal knowledge of appropriate foreign languages. Collectively called toponymists because their occupation focuses on place names, the staff carries out research on names of foreign areas to meet stated requirements or to respond to changes affecting many places. During the Cold War, a principal goal was to provide names for maps and charts of the Soviet Union. Since that country did not distribute maps of its territory at useful scales, the FNC staff had to examine a range of sources to locate names. Similar programs were in place to obtain information for the so-called Iron Curtain countries.

The committee elects a chair and a vice chair for two-year terms. With the formation of the National Imagery and Mapping Agency (NIMA), people formerly from agencies involved with intelligence operations may be more closely involved with FNC functions. NIMA represents the Department of Defense as did its predecessor, the Defense Mapping Agency. The FNC meets approximately each quarter, with additional sessions as required. Occasionally, representatives from other agencies attend meetings to register their interests while people from organizations such as the National Geographic Society as well as retirees may also attend upon request. Including staff members and visitors, the meeting of August 25, 2000, had fifteen participants. The FNC agendas and productivity have always been prodigious. The cited session, its 332d meeting, reviewed one or more names in eight countries. The agenda also identified forty-five countries having issues concerning names that were currently before the staff for ongoing research.

The FNC executive secretary has general responsibility for directing work of the committee's staff, for preparing agendas, circulating reports of meetings, and otherwise assuring proper coordination with relevant NIMA offices so FNC and its staff can meet stated national and international requirements. The official work schedules of the staff, however, are subject to internal NIMA control.

While the committee reviews staff recommendations for new names, it conforms to decisions of the State Department member regarding names of countries. This function relates to that agency's responsibility for U.S. international relations; it is essential that the official name as given by a country be accepted to assure proper political and diplomatic communications. During the past several years, the committee has approved names of some 15,000 places annually. While names of countries or major cities may receive most publicity, the FNC requirement to identify new names of administrative subdivisions (similar to U.S. states and counties) demands much effort. Other changes needing attention also may include the creation of new and the elimination of old administrative units, the subsequent changes of their locations identified by coordinates, and the modification of terms describing their functions. Thus in addition to changing the name of a particular country—which may not be difficult—the FNC may also have to carry out considerable research to trace changes in a sizable number of administrative units in the same country.

The collapse of the Soviet Union and the resulting birth of fifteen independent countries generated an enormous task of tracking new names. The staff at NIMA went into instant overtime, as did offices responsible for revising maps and charts. Many U.S. departments, as well as private publishers, called for updated names information. To its great credit, the staff performed an essential task with promptness and accuracy. Among its other subsequent accomplishments was the depiction of names for maps with complicated ethnic patterns and boundaries which showed areas of the former Yugoslavia that were subject to international briefings as part of the 1995 Dayton accord.

Symbolic of new political conditions developing over the past decade, staff members have also attended meetings of authorities in Ukraine and Lithuania. The advantages are twofold: the committee is able to obtain useful data and the countries are being opened to the principles, policies, and procedures that the committee has developed over half a century.

To the extent possible, the committee also communicates with appropriate authorities in other nations. At the twentieth session of the U.N. Group of Experts on Geographical Names in New York in January 2000, BGN representatives continued their productive liaison with experts from other countries. The most fruitful relationship, one of many years' standing, has been with the U.K. Permanent Committee on Geographical Names for British Official Use. BGN and PCGN representatives meet every two years to discuss common interests and develop collaborative programs. In addition, discussions and exchanges of data occur on virtually a continuous basis. Further, as noted, an increasing number of nations are willing to provide information about their place names. At another level, the Department of State may receive official correspondence from their offices in other countries that gives information about names. Such communications are the basis of many staff recommendations for action by the FNC.

In a radical change of long-standing procedures, the committee staff currently contracts with outside groups to review maps and other documents to identify place names and relevant locational details. The staff then reviews the results and, as required, modifies them for incorporation into recommendations submitted to the committee. This policy allows staff members more time to address other equally important tasks.

Digital techniques have greatly improved the committee's function to disseminate name information. Gazetteers are being digitally produced and access to the Web allows users to obtain lists of names. While printed gazetteers are still available to the public, their stock is diminishing and new editions will be in digital format only. Where specifically required, printouts can be provided. A useful document is the Foreign Names Information Bulletin, a periodic publication that lists decisions of the committee for stated periods.

Below is information from the minutes of typical FNC meeting that illustrates the nature of its function.

Excerpt from Minutes of the 306th Quarterly Meeting of the Foreign Names Committee, January 29, 1997 (comments rewritten to provide full description)

> Agenda item 5.ll. New administrative subdivisions of Slovenia. The principal sources for the new administrative structure of Slovenia were: Republika Slovenja, Občine Januar 1995 (Republic of Slovenia, Communes January 1995) and Zakon o ustanovitvi občin ter o dolo čitvi njihovih območij (Law on the Establishment of Communes and on the Definition of their Territories, 3 October 1994). It was noted that coordinates for the recommended entries were taken from a third source, a smaller scale Slovenian map. The staff prepared a 12-page paper for committee review which noted that 84 new first-order administrative divisions exist now in the place of 1 former first-order and 54 former second-order divisions, and that 6 second-order divisions were abolished.

As published in the minutes of the cited meeting, the format of the cited twelve-page staff paper with its first two entries is below. The format is standard for all FNC staff papers:

FNC 306
Attachment 11
APPROVED

TO: BGN Foreign Names Committee
FROM: Staff
SUBJECT: RECOMMENDED FILE CHANGES: SLOVENIA
DATE: 6 January 1997

Present Entry	Recommended Entry
Ajdovščina, Opstina: ADM2	*Ajdovščina*, Občina: ADM1
46°15′ N., 15°10′ E.	45°54′ N., 13°56′ E.
None	*Beltinci*, Občina: ADM1
	46°37′ N., 16°14′ E.

Note: ADM is the BGN standard code for "administrative entity." ADM1 is a first-order administrative division (like a U.S. state) and ADM2 is a second-order administrative division (like a county in a state). Actions noted above are to upgrade an entity to a higher order, to change the coordinates of its center point, and to respell it. A related action is to form a new entity.

Appendix D

Excerpts from BGN Gazetteers of Undersea Features and Antarctic Names

UNDERSEA FEATURES

The introduction to the 1990 edition of the BGN *Gazetteer of Undersea Features* contains a list of fifty-three terms and their definitions, including nine that apply to smaller features revealed by larger-scale bathymetric charts. Each name in the gazetteer is accompanied by a "designation" term that is generally part of a name but is included to define feature types in cases where the name does not include the appropriate term. When a name in long usage has been replaced by newer nomenclature resulting from improved surveys or evidence that an even earlier name has been in use, the previous name is listed in italicized form. It is followed by a reference to the current name, which is also listed. This double naming permits users to know cases when a name has been replaced by a different name. Below is a section from the cited gazetteer.

Name	Designation	Latitude	Longitude
Hardy Reef	Reef	10° 08′ N	116° 08′ E
Harrie Guyot	Guyot	5° 35′ N	172° 17′ E

Harrington Hill	Hill	31° 21′ N	77° 34′ W
Harrison Seamount	Seamount	12° 40′ N	167° 55′ W
Harris Ridge (see Lomonosov Ridge)	Ridge	88° 00′ N	140° 00′ E

Information on undersea features is available from ACUF staff member Trent Palmer at 301-227-2355.

ANTARCTIC NAMES

The latest gazetteer, *Geographic Names of the Antarctic*, 2d edition, was published in 1995 and contains about 12,000 names (including variant names). Its first twenty pages describe the function of the committee and its naming procedures. Unlike other BGN gazetteers, this publication gives background information on each name. One entry follows:

Armagost, Mount   71°38′S, 166°01′E

One in a series of peaks (2,040 m) that rise between Mirabito Range and Home-run Range in northern Victoria Land. This peak stands 9 mi SW of Mount LeResche. Mapped by the USGS from surveys and U.S. Navy air photos 1960–63. Named by US-ACAN for Chief Equipment Operator Harry M. Armagost, USN, who wintered over at McMurdo Station in 1963 and 1967.

For information about names in Antarctica, contact Roger Payne at 703-648-4544.

Appendix E

Examples of Conventional Names

The U.S. Board on Geographic Names has published several lists of conventional names over the years. The number of sanctioned English conventional names has declined, but some are still in use. Below is an alphabetical selection of some names that remain useful. Notice that some conventionals are versions of names whose local spellings are also in the Roman alphabet, some are translations of foreign terms, and others have been romanized from other scripts. Each name has its conventional form, the local form or forms (as romanized if required), the feature type, and the major area or country (or countries) where the feature is located. The similarity of some local forms and the conventional equivalents can be seen in some cases.[1]

Conventional Name	Local Name	Kind of Feature	Major Area
Anatolia	Anadolu	Region	Turkey
Babylon	Aṭlāl Bābil	Area	Iraq
Cologne	Köln	City	Germany
Damascus	Dimashq	City	Syria
Danube River	Donau	River	Austria and Germany
Danube River	Duna	River	Hungary
Danube River	Dunav	River	Bulgaria
Danube River	Dunărea	River	Romania
Danube River	Dunaj	River	Russia

Mount Kenya	Kirinyaga	Mountain	Kenya
Mocha	Al Mukhâ	City	Yemen
Munich	München	City	Germany
Rhine River	Rhein	River	Germany
Rhine River	Rhin	River	France
Rhine River	Rijn	River	Netherlands

NOTE

1. Names of places in Iraq, Syria, Bulgaria, Russia, and Yemen are romanized from local non-Roman alphabet languages according to BGN/PCGN romanization systems that apply to the writing system of each cited country. Diacritics are used when part of local Roman-alphabet names or when part of transliterated versions of local non-Roman alphabet names.

Appendix F

Selected Place Names from the BGN Digital Gazetteer of Austria

Name	Desig.	Latitude	Longitude	Area	UTM	JOG No.
Törlspitze	PK	47°02′00″N	13°16′00″E	AU02	UN61	NL33–01
Törlspitzen[a]	MT	47°25′00″N	11°08′00″E	AU00	PT65	NL32–03
Tormafalu[b]	PPL	47°47′00″N	16°25′00″E	AU01	XN09	NL33–03
Tormauer	GRGE	47°53′00″N	15°15′00″E	AU03	WP10	NL33–02
Torren	PPL	47°36′00″N	13°09′00″E	AU05	UN67	NL33–01
Torrenerbach	STM	47°36′00″N	13°10′00″E	AU05	UN67	NL33–01
Torscharte	PASS	47°00′00″N	13°32′00″E	AU02	UN80	NL33–01
Torscharte	PASS	47°27′00″N	13°02′00″E	AU05	UN55	NL33–01
Tor See	LK	47°10′00″N	11°41′00″E	AU07	QT02	NL32–03
Tor Spitze	MT	47°11′00″N	11°40′00″E	AU07	QT02	NL32–03

[a]Austria and Germany (feature common to both countries)
[b]See Krensdorf (a variant name)

COLUMN HEADINGS

Name—The name as approved by the Board on Geographic Names and alternate names (if any) for the same feature.

Designation (Desig.)—Two to five letters, occasionally including a number, identify the nature of the feature. The letters generally are abbreviations of the feature type and may derive from the specific element (as translated into English) of the standard name. Designations shown above: GRGE is gorge; LK is lake; MT is mountain; PASS is pass; PK is peak; PPL is populated place; and STM is river or stream. The publication has many other designations defined in the introductory part of the gazetteer.

Latitude and Longitude—The locations of features are defined in degrees, minutes, and sometimes seconds. For features such as cities, administrative divisions, or other areas covering a wide area; lakes; and mountains, center points are used. For rivers or streams, their mouths (where they join a larger body of water) are located.

Area—Two letters and two numbers indicate a feature's location. The letters identify the country according to the code developed by the International Standardization Organization, and two following numbers indicate the relevant administrative division in the country.

UTM—Universal Transverse Mercator, a cartographic projection used by the U.S. National Imagery and Mapping Agency (NIMA) for some of its maps. The number identifies a map of this series where the named feature is located.

JOG No.—JOG stands for a series of NIMA maps called the Joint Operational Graphic. The number identifies the JOG sheet that carries the named feature. The JOGs are at a scale of 1:250,000 and provide coverage for the entire world.

Appendix G

Comparative Examples
of Selected Russian Cyrillic

NAMES ACCORDING TO THE OFFICIAL RUSSIAN SYSTEM
(GOST 1983) AND THE BGN/PCGN SYSTEM

GOST 1983	Russian Cyrillic	BGN/PCGN
Rjazan	Рязань	Ryazan'
Jakutsk	Якуцк	Yakutsk
Ostrov Sahilin	Остров Сахалин	Ostrov Sakhalin (Sakhalin Island)
Holodnyj	Холодный	Kholodnyy
Brjansk	Брянск	Bryansk
Bol'šoj Enisej	Болшой Енисей	Bol'shoy Yenisey
Černigov	Чернигов	Chernigov (Ukraine, now Chernihiv)
Hanty-Mansijsk	Ханты-Мансийск	Khanty-Mansiysk
Vjaz'ma	Вязьма	Vyaz'ma
Idrica	Идрица	Idritsa
Nežin	Нежин	Nezhin

Note: Some names in this column contain diacritics related to their spelling and pronunciation. Information about such diacritics is found in *Romanization Systems and Roman-Script Conventions,* compiled by the foreign names staff of the U.S. Board on Geographic Names.

Appendix H

BGN/PCGN Romanization System for Serbo-Croatian Cyrillic

	Cyrillic				Roman		Cyrillic				Roman
1.	А	а	*А*	*а*	a	16.	Н	н	*Н*	*н*	n
2.	Б	б	*Б*	*б*	b	17.	Љ	љ	*Љ*	*љ*	nj
3.	В	в	*В*	*в*	v	18.	О	о	*О*	*о*	o
4	Г	г	*Г*	*г*	g	19.	П	п	*П*	*п*	p
5.	Д	д	*Д*	*д*	d	20.	Р	р	*Р*	*р*	r
6.	Ђ	ђ	*Ђ*	*ђ*	đ (Đ)[1]	21.	С	с	*С*	*с*	s
7.	Е	е	*Е*	*е*	e	22.	Т	т	*Т*	*т*	t
8.	Ж	ж	*Ж*	*ж*	ž	23.	Ћ	ћ	*Ћ*	*ћ*	ć
9.	З	з	*З*	*з*	z	24.	У	у	*У*	*у*	u
10.	И	и	*И*	*и*	i	25.	Ф	ф	*Ф*	*ф*	f
11.	Ј	ј	*Ј*	*ј*	j	26.	Х	х	*Х*	*х*	h
12.	К	k	*К*	*k*	k	27.	Ц	ц	*Ц*	*ц*	c
13.	Л	л	*Л*	*л*	l	28.	Ч	ч	*Ч*	*ч*	č
14.	Љ	љ	*Љ*	*љ*	lj	29.	Џ	џ	*Џ*	*џ*	dž
15.	М	М	*М*	*М*	m	30.	Ш	ш	*Ш*	*ш*	š

NOTE

This table reflects a BGN/PCGN agreement of 1994 concerning usage of the subject scripts. Serbo-Croatian, the official language of Bosnia and Herzegovina, Croatia, Montenegro, and Serbia, is written in two different standard scripts—Roman and Cyrillic. Croatia uses the Roman script, Montenegro and Serbia use the Cyrillic script, and Bosnia and Herzegovina uses both scripts. The two standard scripts are officially equally equivalent. The italicized letters represent a style of Cyrillic used in certain places. Serbia and Montenegro have asserted the formation of a joint independent nation (Yugoslavia), but this entity has not been recognized as a state by the United States.

　　1. The digraph dj (Dj) will occasionally be found as an alternative form of đ (Đ).

Appendix I

Accessing Information about Foreign and Domestic Place Names Processed by the BGN

ACCESS VIA THE INTERNET

The World Wide Web at the GEOnet names server provides access to the BGN/NIMA foreign geographic names database. The names server contains worldwide geographic names information used to produce BGN gazetteers of foreign countries as well as names information appearing on NIMA maps and charts. The server can be reached through the NIMA homepage at <http://www.nima.mil>. Select the "Maps and Charts Geodata" button to find the link to the GEOnet names server. Links from the GEOnet names server homepage also connect with the Geographic Names Information System of domestic names maintained by the U.S. Geological Survey, with the national database of Canadian place names processed by the Canadian Geographical Names Board, and with a number of other informative Internet sites. This service is provided free of charge.

DIGITAL INTERIM GEOGRAPHIC NAMES DATA ON CD-ROM

A CD-ROM is now available for users requiring a complete copy of the database referenced above. The database is provided in a series of flat files in ASCII format, with diacritics encoded using the scheme employed by NIMA's geographic names processing systems, in a series of hypertext markup language (html) files to allow easy viewing with a Web browser. Additional files are also provided that describe the format and content of the data. No software is included on the CD; users must supply their own application software to display and manipulate the data. The compact disc is available free of charge through U.S. Geological Survey and NIMA distribution channels. The NIMA stock number for this interim product is GAZGNDIGNAMES. Information about ordering the CD from the USGS can be obtained from Debbie Brashear at 703-648-6891.

HARD-COPY GAZETTEERS

Hard-copy BGN gazetteers continue to be available for nearly every country on earth. Inquiries about ordering procedures and pricing should be referred to the U.S. Geological Survey Map Distribution Center in Denver, Colorado, at 303-236-7476.

RECENT CHANGES IN FOREIGN PLACE NAMES

Name changes approved by the BGN Foreign Names Committee can be tracked by subscribing to the *Foreign Names Information Bulletin,* a printed document issued periodically by the committee's staff. The publication is distributed free of charge. To be placed on the mailing list, contact Randall Flynn, BGN executive secretary for foreign names. See reference under "Foreign Names" below.

FOREIGN NAMES INQUIRY SERVICE

Inquiries about foreign names can be submitted to NIMA as the office supporting the staff of the BGN Foreign Names Committee. The service also includes inquiries about names of undersea features, another BGN function that NIMA supports. Questions may be referred to the NIMA Branch of Geographic Names and International Boundaries at 301-227-2360. Requests for such information can also be sent to the same office at 4600 Sangamore Road, Mail Stop D-61, Bethesda, MD 20816-5003

INFORMATION ABOUT STANDARDIZING
FOREIGN AND UNDERSEA NAMES

Questions about such procedures can be sent to

U.S. Board on Geographic Names
Executive Secretary for Foreign Names
NIMA, 4600 Sangamore Road, Mail Stop D-56
Bethesda, MD 20816-5003

INFORMATION ABOUT DOMESTIC AND ANTARCTIC NAMES

As noted above, the GEOnet names server homepage provides a connection with the Geographic Names Information System of domestic and Antarctic names as maintained by the U.S. Geological Survey. Another conduit of names processed by names experts at USGS is <http://mapping.usgs.gov/www/gnis>. Information provided includes names, publications as gazetteers and papers related to BGN work in these areas, and protocols with other nations and organizations regarding work on Antarctic names. For general information, communicate with Roger Payne, BGN executive secretary and executive secretary for domestic names at 703-648-4544; FAX 703-648-4549; or e-mail: rpayne@usgs.gov.

PRINCIPAL PERSONS INVOLVED
WITH BGN NAMES PROGRAMS

Domestic U.S. Names and Antarctic Names

Contact Roger Payne, executive secretary of the U.S. Board on Geographic Names and executive secretary of its Domestic Names Committee at 703-648-4544.

Foreign Names

Contact Randall Flynn, geographer of the National Imagery and Mapping Agency and executive secretary of BGN Foreign Names Committee at 301-227-3050 or Frederic Rohrer, member, NIMA Geographic Names Branch, at 301-227-3059.

Internet Access to Agencies Involved with BGN Programs

The official source at USGS for information about domestic U.S. place names is <http://mapping.usgs.gov/www/gnis>.

The official source at NIMA for information about foreign place names is <http://www.toponym.nima.mil>.

The official source for information about foreign countries at the U.S. Department of State is <http://www.state.gov/www.regions.html>. Then select "Country Information" and "List of Independent States."

Note: The information in this appendix was provided by the USBGN in cooperation with the National Imagery and Mapping Agency and the U.S. Geological Survey. Internet addresses are subject to change. If any of the cited addresses regarding BGN names programs do not respond, users may call the following persons for information. For foreign names call Frederick Roher at 301-227-3059. For domestic names, call Roger Payne at 703-648-4544.

Appendix J

U.S. and Other Sources of Information about Place Names

THE U.S. LIBRARY OF CONGRESS

The Geography and Map Division of the U.S. Library of Congress in Washington, D.C., has a large collection of publications about place names, including many that focus on historical aspects of the topic. Researchers can gain access to its holdings, and division staff identify books available for interlibrary loans. For information, call Ronald Grim at the Geography and Map Division of the LC at 202-707-8532, FAX at 202-707-8531, or e-mail at rgrim@loc.gov.

THE WEB

Some computer search programs produce numerous references to publications and organizations dealing the place names on the search terms "geographic names" or "place names." With the Windows 6.1 program, this action produces a listing of 42,179 topics, most of which deal with names. Included is an item already referred to in Appendix I, the GNS: GEOnet names server, which is also available at <http://164.214.2.59/gns/html>.

THE UNITED NATIONS

The programs of the United Nations to standardize place names have led to the production of a vast library of documents. Summary reports of conferences include references to progress made by member nations and by the Group of Experts, as well as resolutions for ongoing actions. The Group of Experts also issues reports describing work its members have carried out in response either to U.N. conference edicts or to programs inaugurated by its divisions. The resulting library includes summary reports of U.N. meetings, papers submitted by participants, publications such as gazetteers issued by various nations, and accompanying documents. Regrettably, the office supporting U.N. names programs cannot provide free copies of such items to the general public, but the staff can refer callers to a branch that takes orders for some materials, including papers or reports submitted in support of conference agendas. It is essential to identify desired materials as to date, country of origin, subject, and, where possible, conference code numbers of each document. The phone number at U.N. headquarters in New York that deals with place names programs is 212-963-5953, and the FAX is 212-963-0523. The U.N. Group of Experts on Geographical Names *Newsletter* (May 18, 1998) has a list of experts from seventy-two countries along with their titles, addresses, and, where available, telephone, e-mail, and FAX numbers. This is a valuable and "first ever" document of such scope, and persons wanting to know about place names and related publications in subject countries are advised to consult this document. The working group on romanization systems of the UNGEGN has produced a document that shows romanization systems classified as official for U.N. purposes. It is available from a source in Estonia at <http://www.eki.ee/wgrs>. The Geographical Names Board of Canada in Ottawa is assembling a library of U.N. documents, and its staff may be able to provide information. The telephone number there is 613-992-3405. A Canadian report for the 1998 U.N. Conference on Geographical Names identifies the following as a source of information about U.N. documents: <http://geonames.NRCan.gc.ca/english/unindex.html>.

Both of these Web sites identify numerous other sites that provide detailed information about U.N. names programs, dating from the first U.N. conference in 1967 to the most recent event, the twentieth session of the U.N. Group of Experts in January 2000.

Sources Cited

Anderson, David L. ed. *Facing My Lai: Moving Beyond the Massacre.* Lawrence: University Press of Kansas, 1998.

Aurrousseau, Marcel. "Geographical Names for International Use." *Proceedings and Transactions of the Fifth International Congress on Onomastic Sciences,* 1958.

Berkowitz, Freda Pastor. *Popular Titles and Subtitles of Musical Compositions.* 2nd ed. Metuchen, N.J.: Scarecrow, 1975.

Brugioni, Dino. "The Tyuratam Enigma." *Air Force Magazine,* March 1984.

Bryson, Bill. *The Mother Tongue: English and How It Got That Way.* New York: William Morrow, 1990.

Duer, Douglas. Review of Keith Basso, *Wisdom Sits in Places: Language and Landscape among the Western Apache.* Albuquerque: University of New Mexico Press, 1996. In *Professional Geographer* 50, no. 1 (1998). A publication of the Association of American Geographers.

Edson, William D. *Railroad Names.* 3d ed. Potomac, Md: McClain, March 1993. Located in library of the Association of American Railroads, Washington, D.C.

Fischer, David Hackett. *Albion's Seed: Four British Folkways in America.* New York: Oxford University Press, 1989.

Funk, Wilfred. *Word Origins and Their Romantic Stories.* New York: Bell, 1978.

Goodenberger, Jennifer. *Subject Guide to Classical Instrumental Music.* Metuchen, N.J.: Scarecrow, 1989.

Green, Jeff. *The Green Book of Songs by Subject: The Thematic Guide to Popular Music.* Nashville: Professional Desk References, 1995.

Herman, R. D. K. "The Aloha State: Place Names and the Anti-Conquest of Hawaii." In *Annals.* Association of American Geographers, March 1999. Reference to Pukui, M .K., S. H. Elbert, and E. T. Mo'okini. *Place Names of Hawai'i.* 2d ed. Honolulu: University Press of Hawaii, 1974.

Kaplan, Justin, and Anne Bernays. *The Language of Names.* New York: Simon & Schuster, 1997.

Keillor, Garrison. *Lake Wobegon Days.* New York: Viking Penguin, 1985.

MacAoda, Breandan S. "Mineral Names from Toponyms." *Names* 37, no. 1 (1989). Publication of the American Name Society.

Moseley, Christopher, and R. E. Ashler, eds. *Atlas of the World's Languages.* London: Routledge, 1994.

Norris, Kathleen. *Dakota: A Spiritual Geography.* New York: Ticknor & Fields, 1993.

Orth, Donald J. "The Story of the Naming of Mt. Rainier and Other Domestic Names Activities of the U.S. Board on Geographic Names." *Names* 32, no. 4 (1984).

Pillar, Ingrid. *American Automobile Names.* Essen, Germany: Verlag Die Blaue Eule, 1996.

———. "Variation in Automobile Naming." *Names* 47, no. 2 (1999). American Name Society, June 1999.

Proust, Marcel. *Remembrance of Things Past: Swann's Way.* New York: Random House, 1934.

———. *Remembrance of Things Past: The Guermantes Way.* New York: Random House, 1934.

Ray, Angela G. "Calling the Dog: The Sources of AKC Breed Names." *Names,* 43, no. 1 (1995).

Reed, W. L., and M. J. Bristow, eds. *National Anthems of the World.* 8th ed. London: Cassell PLC, 1993.

Romanization Systems and Roman-Script Conventions. Bethesda, Md.: United States Defense Mapping Agency, 1994.

Shipley, Joseph T. *Dictionary of Word Origins.* 2d ed. New York: Philosophical Library, 1945.

"Soviet Cartographic Falsification." *Military Engineer* 62, no. 410 (1970). Publication of the American Society of Military Engineers.

Stewart, George R. *Names on the Land.* New York: Random House, 1972. First published in 1945.

———. *Names on the Globe.* New York: Oxford University Press, 1975

Szarkowski, John. *The Porfolios of Ansel Adams.* Boston: Little, Brown, 1981.

"The National Gazetteer of the United States of America." Concise ed., in U.S. Geological Survey *Professional Paper 1002-US,* December 1984.

The World Factbook. Washington, D.C.: Central Intelligence Agency, 1999.

Times Atlas of the World: Comprehensive Edition. London: Time Books, 1992.

Zelinsky, Wilbur. *Exploring the Beloved Country: Geographic Forays into American Society and Culture.* Iowa City: University of Iowa Press, 1994.

Useful Published Sources

SOURCES CITED IN ENDNOTES

Annals of the Association of American Geographers. A publication of the Association of American Geographers, 1710 16th Street, N.W., Washington, D.C. 20005. Occasional articles relate to place names.

Merriam Webster's Geographical Dictionary. 3d ed. Springfield, Mass.: Merriam-Webster, 1997. An excellent collection of place names and associated information.

Names. A publication of the American Name Society. This quarterly journal is the foremost publication concerning place names. Although it focuses on topics such as personal and literary names, virtually every issue deals with place names either as a specific article or as a book review. Five issues between 1969 and 1990 had bibliographies on place name literature, mostly American but also Canadian. The entire collection of *Names* from the first edition in 1954 is housed in the Coltharp Collection at the University of Texas in El Paso. Persons interested in the contents of the publications and their availability can obtain information by calling Roberta Arney at 915-747-6702 or communicating via e-mail at <http://libraryweb.utep.edu/ref/onomast.html>. Information about subscriptions is available from Edward Callary, editor, English Department, Northern Illinois University, DeKalb, IL, 60115; e-mail: ecallary@niu.edu.

Nelson, Derek. *Off the Map: The Curious Histories of Place-Names.* New York: Kodansha International, 1997. A collection of very interesting histories of a number of place names, some of them well known and some arcane. The book also tells about efforts to deal with names in areas having different languages and cultural situations.

Shipley, Joseph T. *Dictionary of Word Origins.* 2d. ed. New York: Philosophical Library, 1945. This venerable publication contains an alphabetic listing of words, including those whose origins are derived from place names.

Stewart, George R. *Names on the Land*. New York: Random House, 1945. This is a classic tome on the subject of place names. Subsequent editions up to 1975 are of the same ilk.

OTHER SOURCES RELATED TO PLACE NAMES

Davis, Kenneth C. *Don't Know Much about Geography*. New York: W. Morrow, 1992. A book covering many aspects of the field of "geography," from historically factual to whimsical, this volume also gives background on origins of American state names. It lists countries and their dates of origin by continent.

Dunkling, Leslie Alan. *The Guinness Book of Names*. Enfield, U.K.: Guinness, 1989. The book covers interesting and unusual attributes of names of all types, including place names. For example, it gives the origins and nicknames of U.S. states as well as long and short names of the world.

Efvergren, Carl. *Names of Places in a Transferred Sense in English*. Lund, Sweden: Håkan Ohlsson, 1909. Reprint; Detroit: Gale Research Company, Book Tower, 1969. This early work provides background on the factors causing place names to acquire secondary or even tertiary meanings. The book describes some 500 such words listed in nearly twenty categories as animals, food, cloth, and also as parts of common expressions having humorous or derogatory slants. The work refers to about fifty publications, some of which date to the 1860s. Many of the words cited as thus being transferred are no longer in common use. The book also notes that, except for English, few if any other languages have a vocabulary containing so many words of this nature.

Jouris, David. *All Over the Map: An Extrordinary Atlas of the United States*. Berkeley, Calif.: Ten Speed, 1994. This book presents thirty-three maps of the United States, showing towns whose names fall into categories such as sports, religion, occupations, animals, and music. The author also gives brief backgrounds on the origins of names. While designed to evoke humorous reactions, the book nevertheless provides an interesting perspective on U.S. place names.

Room, Adrian. *Place-Name Changes, 1900–1991*. Metuchen, N.J.: Scarecrow, 1993.

_____. *Place-Name Changes since 1900: A World Gazetteer*. Metuchen, N.J.: Scarecrow, 1979.

Note: The author has consulted numerous publications, including books, articles in professional journals, and materials produced by the United States, the United Nations, and other official national or international organizations. This list includes some of the documents the author used, as well as additional items. As noted, some sources are located in libraries of federal agencies and may not be easily available. Many notes from such items were extracted at various times during the author's employment with the U.S. Board on Geographic Names and the Defense Mapping Agency. In addition, the author has copies of documents long out of print yet relevant to the story. Subscriptions to some publications are possible, as noted.

Index

The index does not carry all of the many place names included in the book principally as examples of names having relationships with noted topics.

About the Author

Dr. Richard R. Randall's career of many years in various professions has given him a comprehensive understanding of place names in their broadest context. He studied civil engineering, geography, and received a Ph.D. in political geography after conducting research in Austria as a Fulbright Scholar. He worked as a geographer with the Central Intelligence Agency, as the Washington representative for Rand McNally and Company, and as the chief geographer for the Defense Mapping Agency, during which assignment he also served as the executive secretary of the interagency U.S. Board on Geographic Names and its Foreign Names Committee. The latter assignment of twenty years permitted Dr. Randall to apply his broad knowledge of geography, cartography, and foreign languages to the board's task of providing accurate place names to support official U.S. programs on a global basis. In this position, he administered the work of a staff of some fifty names experts at the U.S. Geological Survey and the Defense Mapping Agency and collaborated with members of the board who represented nine U.S. agencies. This position gave him a keen understanding of procedures regarding the treatment of place names to meet stated requirements. In addition, he played significant roles in the development and implementation of U.N. programs to standardize names for international purposes. For several years in Latin American countries, he taught two-week courses on methods to collect and record place names based on a curriculum he created. He is a member of professional organizations in surveying and mapping, geography, and place names and is listed in the current edition of "Who's Who in America." Since retiring in 1993, Dr. Randall has continued to attend BGN and U.N. meetings and to collaborate with professional colleagues in the United States and abroad. His long involvement with U.S. and international names programs, along with his strong interests in music and literature, have given him unique qualifications to author a book that reveals the important role place names play in defining the world—and more.

CPSIA information can be obtained
at www.ICGtesting.com
Printed in the USA
LVOW11*1747180617

538521LV00017B/576/P

9 780810 839069